Sound Control in Buildings

Sound Control in Buildings

A Guide to Part E of the Building Regulations

M.W. Simons
BSc, MSc, PhD, MIOA, MCIOB

and

J.R. Waters
BSc, MPhil, PhD, MCIBSE, CEng

Blackwell
Publishing

Editorial offices:
Blackwell Publishing Ltd, 9600 Garsington Road, Oxford OX4 2DQ, UK
 Tel: +44 (0)1865 776868
Blackwell Publishing Inc., 350 Main Street, Malden, MA 02148-5020, USA
 Tel: +1 781 388 8250
Blackwell Publishing Asia Pty Ltd, 550 Swanston Street, Carlton, Victoria 3053, Australia
 Tel: +61 (0)3 8359 1011

First published 2004 by Blackwell Publishing Ltd

Library of Congress Cataloging-in-Publication Data
is available

ISBN 1-4051-1883-0

A catalogue record for this title is available from the British Library

Set in 10/12.5pt Times
by DP Photosetting, Aylesbury, Bucks
Printed and bound in the UK using acid-free paper
by TJ International, Padstow, Cornwall

For further information, visit our website:
www.thatconstructionsite.com

Contents

Preface

Unwanted noise is a disturbing feature of many aspects of modern life. It is often a significant nuisance within buildings, and domestic dwellings are particularly vulnerable to noise transmitted from attached properties. Part E of the Building Regulations for England and Wales sets minimum requirements for the resistance to the passage of sound, and the new edition that came into force on 1 July 2003 introduced substantially improved standards and stringent new requirements for the testing of completed buildings.

There are two consequences. Firstly, separating walls and floors must not only have significantly improved design detailing in themselves, but also be designed to include the sound transmission effects of all adjoining wall and floor components. Secondly, because the standard of workmanship in the construction of a building is at least as important in reducing noise as the design, the only certain criterion for compliance with the regulations is that the completed building should pass a sound transmission test. It is the responsibility of the builder to ensure and demonstrate that compliance has been achieved, and the consequences of failure can be severe. Significant remedial work may be required not only to the failed dwelling but also to any other dwelling with similar construction details. It is therefore necessary for both designers and builders to understand the principles of design for resistance to the passage of sound, and the necessity for observing correct and sufficient procedures in the construction process.

This book describes and explains those aspects of the resistance to the passage of sound in buildings that are essential to an understanding of Approved Document E (2003 edition). The aim throughout is for the book to be a companion and guide to Approved Document E, and other relevant documents, and not a substitute for them. Part 1 is a general introduction to the basic background theory of sound absorption, airborne sound transmission, impact sound transmission, and vibration isolation, leading to an explanation of the principles underlying the design of sound insulating construction in dwellings and other buildings. Part 2 of the book deals with Part E of the regulations and Approved Document E in detail. The regulations and their requirements are described, and the role of testing and the associated test procedures are explained in detail. The recommendations for good design and construction are described, and the role of proposed robust standard details is explained.

The book is intended to be of particular value to all who are involved in the production of dwelling spaces, including architects, designers, surveyors, builders, testing organisations, and also all those in building control who must ensure that compliance with Part E has been achieved.

October 2003

Acknowledgements

The authors are grateful to the following for permission to reproduce copyright material.

Material in Approved Document E (2003 edition) is crown copyright and is reproduced with the permission of the Controller of HMSO and the Queen's Printer for Scotland.

Extracts from BS EN ISO 717-1 and 717-2: 1997 are reproduced with the permission of the British Standards Institution under licence number 2003SK/0121. Extracts from BS EN 150 11654: 1997 are reproduced with the permission of the British Standards Institution under licence number 2003SK/036. BSI publications can be obtained from BSI Customer Services, 389 Chiswick High Road, London W4 4AL, United Kingdom. (Tel +44 (0)20 8996 9001). Email: cservices@bsi-global.com

The authors would also like to thank their wives, Kate and Patricia, for their encouragement and forbearance during the preparation of this book.

PART 1
BACKGROUND ACOUSTICS

Part 1 is an introduction to the acoustics of sound insulation in buildings. The purpose is to provide a framework for understanding the purpose and requirements of the Building Regulations with respect to the resistance to the passage of sound. This is not a textbook on acoustics, but rather a selection of those topics which are most relevant to an understanding of the material to be found in Part 2. The treatment is intended to be at a level suitable for those who must understand and apply the regulations but who otherwise do not have a background knowledge of acoustics. For a more advanced coverage of the subject, the reader is referred to one of the many comprehensive texts that are available.

The material is presented in a logical order, starting with sound absorption and then progressing to airborne and impact sound insulation and vibration isolation. However, it is recognised that many readers will use this book to look up specific topics rather than reading through the chapters in sequence. To accommodate this, there is some overlap and repetition of material between Parts 1 and 2. It should also be noted that British and International Standards are not given reference numbers in the text, but are collected together in alpha-numeric order in the reference section

1 Introduction and Background to the Regulations

1.1 Noise sources and the purpose of the regulations

Part E of the Building Regulations is concerned with sound reduction, and in common with all other parts of the Building Regulations, it has been expanded and revised at regular intervals. The requirements of Part E to the Building Regulations 2000 (as amended by statutory instrument SI 2002/2871) are dealt with in Approved Document E (2003 edition) [1], which came into effect on 1 July 2003. The regulation itself covers sound reduction within and between dwellings, and it also covers acoustic conditions within schools. However, the Approved Document E (2003 edition), although it includes a statement of the full legal requirements of the regulation, deals exclusively with methods of meeting the requirement for dwellings. Methods for schools are to be found in a separate publication on school design [2]. This book is concerned solely with the legal requirements and methods of compliance for dwellings, as contained in the Approved Document E (2003 edition) and referred to in this book as AD E.

The main purpose of the AD E, as stated in the consultation document of January 2001 [3], is to raise sound insulation standards in order to address the problems of noise in dwellings and other rooms used for residential purposes. The reason is evidence that the noise nuisance suffered by the occupants of dwellings has been steadily increasing, with complaints about domestic noise trebling in the ten-year period up to 1998. Several organisations have independently noted increases in complaints of noise disturbance in dwellings, including the Chartered Institute of Environmental Health (CIEH) [4], the English House Condition Survey (EHCS) [5], the National House-Building Council (NHBC) [6], and the Building Research Establishment (BRE) [7]. The 1997/8 CIEH survey reported 148 006 complaints of domestic noise, a figure which has continued to rise since. According to the EHCS, noise problems are independent of the age of the property, with new dwellings as likely to be the source of complaints as old ones. This is consistent with the fact that the NHBC regularly receives complaints of poor sound insulation in new properties. The most frequently complained about dwellings are flats, followed by terraced, then semi-detached, and finally detached houses. This indicates that a major part of the problem is inadequate sound insulation in separating walls and floors.

Besides causing annoyance, unwanted noise can have adverse health effects. The most obvious of these is disturbance of rest and sleep, but a Medical Research Council report [8] lists a total of twelve different possible adverse effects attributable to noise in dwellings. In the case of very high noise levels, it is possible to quantify the damage to hearing caused by particular levels of noise. However, in dwellings, unwanted noise is scarcely ever high enough to produce measurable hearing loss, and so it is difficult to relate noise annoyance and other adverse effects to a specific level of the noise that is causing it. Nevertheless, the consultation document [3] provides estimates of the improvements in occupant satisfaction that may be expected to accrue from an improvement in the sound insulation performance of building elements.

Other factors that were taken into account in framing AD E include changes to the occupancy and usage of the space within dwellings, and changes to the sources of noise. In terms of occupancy and usage, it is now common for rooms to be multi-purpose. For example, a room originally designed as a bedroom may also be used as living space, with television and audio equipment, or used as a study. It is also increasingly common for dwellings to be equipped with more than one bathroom, either independently or en-suite to a bedroom. Both of these developments have implications for noise transmission to other rooms in the same dwelling. The equipment used in the home can be particularly noisy, especially sound reproduction equipment, which is often capable of producing a large amount of acoustic energy at the low frequencies at which the sound insulation performance of building elements tends to be weakest.

In addition to the above, AD E attempts to take into account a general expectation that there should be an upward trend in living standards. Thus it may be argued that the preceding standard was set too near to the threshold at which the noise of normal activities from an adjoining dwelling could be heard. That is, the preceding standard was based on the assumption that such noises, when generated at a reasonable sound level, are just audible. AD E attempts to ensure that normal everyday sounds from a neighbouring dwelling will be inaudible. However, complete insulation against unusual or unreasonable sound is not practicable and is not intended, and in these circumstances protection or redress must be sought outside the Building Regulations.

The improvement in the standards set by AD E with respect to the preceding standards has been achieved in several ways:

- The specific requirements limiting airborne and impact sound transmission have been replaced by a single more general one in Requirement E1
- The Requirement E1 now includes institutional accommodation such as hostels and student halls of residence
- The sound insulation index used to demonstrate compliance with E1 has been modified to place greater emphasis on low frequency sound
- In order to ensure high standards of construction, a system of pre-completion testing for separating elements has been introduced
- Requirements have been introduced for sound insulation between certain rooms within a dwelling

- Requirements have been introduced to control the reverberation of sound in the common internal parts of buildings which contain more than one residential unit
- A requirement has been introduced to control acoustic conditions in schools.

1.2 The relevant legislation etc.

The legislation and associated documents covering the design and construction of buildings are extensive. As far as the resistance to the passage of sound in dwellings is concerned, the most important legislative instruments are:

- Schedule 1 to, and Regulations 5, 6, 7 and 20A of:
  The Building Regulations 2000 (SI 2000/2531) for England and Wales as amended by:
      The Building (Amendment) Regulations 2001 (SI 2001/3335) and
      The Building (Amendment) Regulations 2002 (SI 2002/440) and
      The Building (Amendment) (No. 2) Regulations 2002 (SI 2002/2871).
- Regulation 12A of:
  The Building (Approved Inspectors etc.) Regulations 2000 as amended by:
      SI 2002/2872.

These instruments are legally binding. Schedule 1 states the requirements that must be met with respect to sound insulation and acoustic conditions. Regulations 5 and 6 are concerned with the meaning of, and requirements relating to, material change of use. Regulation 7 lays down a requirement that the materials must be adequate and proper for their purpose, and that building work must be carried out in a workmanlike manner. Regulations 20A and 12A refer to sound insulation testing, and place upon the building control body, whether it be a local authority (regulation 20A) or some other approved building inspector (regulation 12A), the duty to ensure that appropriate sound insulation testing is carried out on completed buildings.

In addition to the legislative instruments there are a number of documents and standards offering practical guidance on design and construction techniques for meeting the legislative requirements. By far the most important of these is the Approved Document E (AD E). The status of AD E is intended to be advisory, and indeed it is clearly stated within it that there is no obligation to adopt any particular solution contained in an Approved Document if the designer and builder prefer to meet the relevant legal requirement in some other way.

Nevertheless, AD E not only provides guidance on building design and construction details, but also gives a number of specific targets and specifications. These include a set of precise target values for airborne and impact sound insulation. It would be difficult to argue, except in the most exceptional circumstances, that the legal requirement could be met with less stringent values of these parameters. Furthermore, the Approved Document gives guidance on appropriate test programmes and specifications for procedures for sound insulation testing. Regulations 20A and 12A require that (where applicable) sound insulation testing should be carried out using a procedure approved by the Secretary of State. The approved testing procedure is Appendix B of AD E. It is therefore reasonable to consider AD E as having greater significance than mere guidance, and in some respects it has a status as important as the legislative instruments it serves.

1.3 Frequency distribution and the response of the human ear

The human ear is capable of detecting sound over a wide range of frequencies. In ideal listening conditions, and for a healthy young adult with normal hearing, the lower limit is around 50Hz and the upper limit approaches 20kHz. However the sensitivity of the ear varies substantially over this range, as shown by the equal loudness contours in Fig. 1.1.

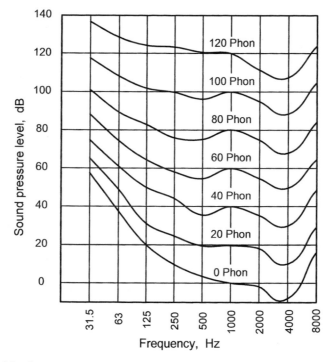

Fig. 1.1 Equal loudness contours.

In this diagram, each contour is constructed by taking a reference sound source with a frequency of 1kHz and a set sound pressure level (say 40dB for the 40dB contour). A second sound source is then operated simultaneously at different frequencies, and at each frequency its sound pressure level is adjusted until it sounds equally loud to the reference sound. Each contour then has a constant loudness, the unit of loudness being the phon, and the value is taken to be the sound pressure level of that contour at the reference frequency of 1kHz. It is clear from the diagram that as the frequency is reduced below the 1kHz reference, more and more sound energy is required to achieve the same impression of loudness. Thus a loudness of 40 phon at 100Hz requires a sound pressure level of about 52dB. Above 1kHz, the contours vary before climbing steeply above 15kHz. These equal loudness contours can be used to set the frequency response of sound level meters so that they imitate the frequency response of the human ear. As the contours vary in shape, it is necessary to choose the most appropriate. It is normal practice to choose three, the 40, 70 and 120 phon contours, called respectively the A, B and C weighting networks, of which the most commonly used is the A network, and sound pressure levels measured using this are designated dBA.

The apparent relevance of the equal loudness contours to sound insulation in buildings is that more insulation is needed at high frequencies than at low frequencies. Whilst this may be true in general terms, there are several other factors to be considered, such as:

- The equal loudness contours were obtained under ideal conditions using only two pure frequencies; normal noises in a domestic environment are far more complex
- The sound pressure level of low frequency sounds from some types of domestic equipment may be very high, suggesting that insulation at low frequencies may be important
- The perception of a noise, and the degree of nuisance it generates, depend on a wide range of factors, and not solely on its frequency and sound pressure level.

1.4   Sound levels and frequency characteristics of noise sources

The sound levels produced by normal speech are usually in the range 40 to 50dBA at a distance of 2 m in front of the speaker. Domestic equipment other than television sets or audio equipment can generate up to 70dBA, though most are much quieter than that. Television sets, and more especially audio equipment, are often capable of sound levels above 85dBA, which is entering the region where hearing loss or even damage may occur.

Although the dBA level of a noise source is a good guide to its noisiness, it is necessary to know the frequency content of the noise if its effect is to be fully understood. The majority of the noises of relevance to sound insulation in dwellings are essentially 'broad band', which is to say that they contain frequencies across the full range of the audible spectrum. In general, the sound energy of a broad band noise is not constant with frequency, and sometimes varies in an unpredictable fashion. To take account of this variation, it is usual to measure the sound energy or the sound pressure level in bands of frequencies. The bands may be of different width:

- Octave bands   the largest bandwidth that is normally used
- Third octave bands   give more detail than octave bands, and are the preferred bandwidth in sound transmission

- Twelfth octave bands   give very fine detail, and are rarely used in building acoustics.

In general, the acoustic energy of domestic noise sources is evenly distributed across the audio spectrum, with a tendency for most of the energy to be at the lower end of the frequency range, that is, below about 2kHz. This is one of the reasons for setting an upper frequency limit to building acoustics measurements. There are of course some exceptions to the general rule; for example an electric drill, when used to drill into masonry, may create significant acoustic energy at high frequencies. Some sources have a prominent single tone or frequency that increases their audibility in the presence of other noises and makes them easily recognisable (e.g. telephones, door bells, alarm clocks, etc.). The most likely source of domestic noise at very low frequencies, below say 200Hz, is audio equipment such as hi-fi systems and electric musical instruments. These often have an artificially boosted bass output to satisfy the demands of the user.

1.5   Noise nuisance

The nuisance, annoyance and disturbance caused by noise depend on many factors. Apart from the characteristics of the noise itself, there are factors relating to the noise sensitivity of the listener and the circumstances in which the noise is heard. Many of these factors cannot easily be quantified, except for very specific situations. For example, the effect of background noise on speech intelligibility is well established. In general, a list of noise nuisance factors may include at least the following:

- The sound pressure level of the noise
- The frequency distribution of the noise
- The presence or otherwise of a particular characteristic in the noise such as a prominent pure tone
- Whether the noise is continuous or intermittent, and if the latter, whether or not it has a regular pattern of intermittency
- The time of day at which the noise occurs
- The circumstances in which the noise is heard.

To add to the difficulty of assessing noise nuisance, many noises that are pleasurable and sought after in some circumstances become a nuisance to the same person when heard in a different context. And also, a particular noise that delights one person may be anathema to another.

The Regulations and methods of demonstrating compliance described in AD E do not attempt to deal with noise nuisance directly. However, the insulation standards in AD E are based on reports and surveys of noise complaints by the occupants of dwellings. The expectation, therefore, is that the application of the 2003 edition of AD E will lead to significantly reduced noise nuisance in dwellings.

The section that applies to schools, that is Regulation E4, has a slightly different objective. In this case, although the reduction of noise nuisance is intended, the main purpose is to enable a school to function properly. This includes a consideration of the ingress of externally generated noise, and also the design of rooms to be acoustically appropriate for their purpose.

1.6 Noise control principles for dwelling-houses, flats and rooms for residential purposes

There are many possible paths by which the sound could travel from one habitable space to another, the main ones being:

- Directly through a separating (party) wall and/or through a separating floor/ceiling
- Indirectly due to vibration of the building structure and/or via small gaps and weaknesses at the junctions of building elements and where electrical and piped or ducted services are mounted (flanking transmission)
- Via the piped or ducted services themselves
- Via external air paths due, say, to openings (doors or windows) being too close together.

The principles and factors governing the control of these sound transmission paths may be summarised as follows.

1.6.1 Direct sound transmission

The factors which contribute to resistance to direct sound are:

- The mass of the separating element. In general, the greater the mass the better the resistance to sound transmission
- The number of independent layers. In general two or more independent (i.e. not physically connected) layers will have a better resistance to sound transmission than a single layer of the same overall mass
- The continuity of the layer or layers. Even a very small break can provide a sound transmission path that is sufficient to create a significant reduction in the resistance to the transmission of sound.

AD E gives detailed guidance on the design and construction of separating elements.

1.6.2 Flanking transmission

Flanking transmission can occur in several different ways. The acoustic vibrations set up in the separating element can transmit to all parts of the building structure. Likewise, acoustic vibrations set up in elements other than the separating element can transmit to the rest of the building structure, including a separating element.

Once these vibrations have entered the structure, they can re-emerge as sound in any other part of the building. Flanking transmission can also occur around the edges of separating elements if the edge details and the junctions with other elements are not adequately designed and constructed. The exact mechanisms by which flanking transmission occurs are often very complex, making flanking transmission particularly difficult to control. Consequently a substantial part of the guidance in AD E is concerned with the design of junctions between separating elements and other parts of the building structure in order to minimise flanking transmission.

1.6.3   Building services

In the case of dwelling-houses, there is usually no connection between the services in one house and any adjoining house or building. If there is no such connection, then provided noise generating appliances such as fans, fan-assisted boilers, pumps, etc. are not mounted on a separating wall, there should not be any special need for noise control measures. However, in flats and rooms used for residential purposes there is often commonality in the supply of services systems, and in these cases considerable care is needed to avoid creating sound transmission paths between residential spaces. Both pipework and ductwork can be efficient transmitters of sound, but AD E gives little guidance on this matter (see Chapter 11, section 11.1 of this book and BS 8233:1999).

1.6.4   External noise paths

In addition to sound transmission through the elements of the building, noise nuisance can arise due to sound transmitted externally, for example through windows and doors. This can be minimised by paying attention to the separation and positioning of windows and doors. For example, the windows on either side of the separating wall between a pair of semi-detached houses must not be placed too close together. Similarly, in a residential building, a reasonable separation between windows must be maintained. Again, AD E gives no guidance on this matter.

1.6.5   Separation of noisy areas from noise sensitive areas

In addition to a consideration of the paths by which sound may be transmitted from one space to another, it is good practice to avoid a building layout in which noisy areas are adjacent to noise sensitive areas. In the case of dwellings, the most noise sensitive areas are likely to be bedrooms, and so the positioning of bedrooms in the overall layout must be carefully considered. Again AD E provides little guidance on this matter (see Chapter 11, section 11.1, BS 8233:1999 and *Sound control for homes* [12]).

2 Sound Absorption and Room Acoustics

2.1 Definition of sound absorption and sound absorbing materials

When a sound wave strikes a surface, some of the energy is reflected, and the rest is absorbed and transmitted, as shown in Fig. 2.1. The sound absorption coefficient, α, is defined as:

$$\alpha = \frac{\text{Sound energy absorbed and transmitted per unit area}}{\text{Sound energy incident per unit area}}$$

Note that because the absorbed and transmitted sound energy are lumped together in the definition, the absorption coefficient is, in effect, a measure of the sound energy that is *not* reflected. The sound reflection coefficient, ρ, is therefore given by:

$$\rho = 1 - \alpha$$

The absorption coefficient depends on:

- The angle of incidence of the sound wave on the surface
- The frequency of the sound wave.

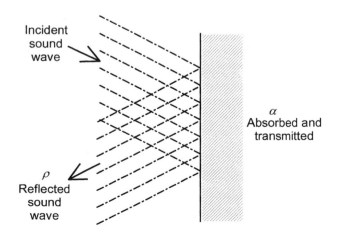

Fig. 2.1 Sound absorption at a surface.

Two types of measurement are made. The normal incidence sound absorption coefficient is obtained when a plane sound strikes the surface at right angles. It requires only small samples of a material to be available. The diffuse sound absorption coefficient is measured when the sound wave is completely random in direction. It is usually measured on large samples in a diffuse sound field in a reverberation chamber, which gives an average value over all possible angles of incidence. Most materials absorb more sound at normal incidence, and soft surfaces normally absorb more sound than hard surfaces. The diffuse absorption coefficient is more appropriate to the sound field within rooms, and so in room acoustics the diffuse absorption coefficient is always used.

Although there is considerable variation in the sound absorbing properties of materials, four main types, shown in Fig. 2.2, can be identified:

- Solid non-porous materials
- Porous absorbers
- Panel (or membrane) absorbers
- Cavity absorbers.

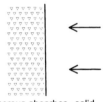

Non-porous absorber - solid
material with a hard surface

Porous absorber - soft material
with an open porous surface

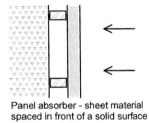

Panel absorber - sheet material
spaced in front of a solid surface

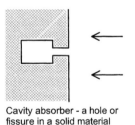

Cavity absorber - a hole or
fissure in a solid material

Fig. 2.2   Types of sound absorbing material.

Figure 2.3 shows the variation of sound absorption coefficient with frequency for examples of each type. Typically, solid materials and porous absorbers show a steady increase in the absorption coefficient as the frequency increases. Masonry walls, either plain or plastered, are typical examples of the former, and any material with an open cellular or fibrous structure, for example mineral wool quilt, is an example of the latter.

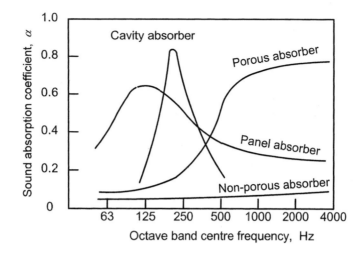

Fig. 2.3 Characteristic frequency absorption curves.

Panel absorbers normally have a peak in the absorption–frequency curve, usually at low frequency, followed by a steady decline as frequency increases. Plasterboard, plywood, or any similar sheet material mounted in front of a solid surface is an example. The frequency at which the peak occurs can be altered by changing the spacing between the panel and the surface behind it.

Cavity absorbers have a sharp absorption peak at a frequency determined by the dimensions of the cavity. The larger the dimensions of the cavity, the lower is the frequency at which the peak occurs. Cavity absorbers are sometimes individually designed and made to absorb at a specific frequency, but they are more often created by forming perforations in a sheet material.

The fact that porous absorbers are better at absorbing high frequency, and panel and cavity absorbers have the ability to absorb at low frequency, has led to the development of composite constructions which combine some of the characteristics of all three. Such is the complexity of the acoustic behaviour of composite constructions that it is possible to predict their performance only in general terms. Reliable values of absorption coefficients must be obtained by measurement, and a range of practical examples is given by Lord and Templeton [9]. Table 12.4 in Chapter 12, which is taken from AD E, lists the diffuse sound absorption coefficients of some common building materials.

2.2   Reverberation of sound in a room

The sound waves in an enclosure must strike the surrounding surfaces, where they are partly reflected and partly absorbed. Any enclosure may contain several surfaces of different area and absorption coefficient. If the areas of the surfaces are denoted by S_1, S_2, S_3 etc. and their absorption coefficients by α_1, α_2, α_3 etc. then:

Area of all surfaces $\qquad S_t = S_1 + S_2 + S_3 + \ldots$

Sound absorption of all surfaces $\quad = \alpha_1 S_1 + \alpha_2 S_2 + \alpha_3 S_3 + \ldots = \sum_{i=1}^{n} \alpha S$

where i is an integer and n is the total number of surfaces.

The average sound absorption coefficient of the surfaces, $\bar{\alpha}$, is sometimes useful, and is defined by:

$$\bar{\alpha} = \frac{\displaystyle\sum_{i=1}^{n} \alpha S}{S_t}$$

The total sound absorption in the enclosure, A_t, is found by adding to the surface absorption any contributions due to the room contents (people, furniture, the air, etc.). If A_c is the total of this additional sound absorption, then

Total sound absorption $\quad A_t = \displaystyle\sum_{i=1}^{n} \alpha S + A_c$

The total sound absorption determines the reverberation of sound in a room. The reverberation, or 'echo', decreases as the absorption increases. The measure of reverberation is the reverberation time, defined as the time it takes for a sound to decay by 60 dB after it is switched off. The reverberation time, T, may be found from the Sabine equation:

$$T = \frac{0.16V}{A_t}$$

where V is the volume of the room. This formula can lead to noticeable errors when the absorption in the room is large (usually when $\bar{\alpha}$ is greater than 0.3). If this is the case and better accuracy is needed, the Norris-Eyring formula may be used:

$$T = \frac{0.16V}{S_t \log_e(1 - \bar{\alpha}) + A_c}$$

Otherwise the Sabine formula is sufficiently accurate for most cases.

2.3   Typical reverberation times

The fact that absorption coefficients vary with frequency means that the reverberation time of a room will also vary with frequency. Typically, the reverberation time will decrease as the frequency increases, and it has been found that a moderate decrease with frequency is preferred by most listeners. Reverberation time also tends to decrease as the size of a room decreases, so that the rooms in dwellings have shorter reverberation times than large halls.

The optimum reverberation time for a room depends partly on its size and partly on its function. Too much reverberation causes sounds to become merged, and in particular reduces the intelligibility of speech. On the other hand, some reverberation is necessary to assist in the diffusion and propagation of sound in a room, especially for music, for which some reverberation assists in 'blending' the sounds from different sources. Because the reverberation time changes with frequency, it is common practice to quote optimum values at a frequency of 500 Hz. These are tabulated, or more commonly presented diagrammatically as in Fig. 2.4, for many typical rooms, but they may also be estimated from the formula due to Stephens and Bate [10]:

$$T_{500} = k[0.0118\left(\sqrt[3]{V}\right) + 0.107]$$

where T_{500} is the optimum reverberation time at 500 Hz
      k is a constant which depends on the main function of the room
      V is the volume of the room, m³.

The constant k is 4 for speech, 5 for orchestral music and 6 for choral music. For a typical living room, and assuming listening to speech is the most frequent function, the formula gives an optimum reverberation of between 0.5 and 0.6 s.

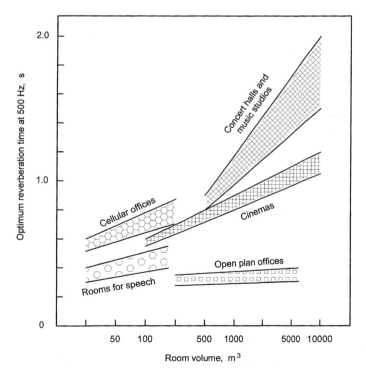

Fig. 2.4 Optimum reverberation times at 500Hz.

2.4   Sound levels in rooms

The sound pressure level, L_{dir}, received directly from a small source of sound that is radiating uniformly in all directions obeys the inverse square law, and is given by the equation:

$$L_{dir} = 10\log_{10} W - 20\log_{10} r - 11 \quad dB$$

where W is the acoustic power of the source
r is the distance of the listener from the source.

If the source is not radiating uniformly, then it is necessary to allow for the non-uniformity by means of a directivity factor, Q, which is defined by the ratio:

$$Q = \frac{I}{I_{av}}$$

where I is the intensity at a point in a given direction from the source
I_{av} is the intensity that would have existed at the same point if the source had been uniform.

The direct sound pressure level for a non-uniform source may then be written:

$$L_{dir} = 10\log_{10} W - 20\log_{10} r - 11 + 10\log_{10} Q \quad dB$$

When the source is situated within a room, there is an additional sound pressure level due to the reverberant sound, that is, the sound reflected back from the surrounding surfaces. The reverberant sound pressure level, L_{rev}, is given by:

$$L_{rev} = 10\log_{10} W + 6 - 10\log R_m \quad dB$$

where R_m is the room constant.

The value of the room constant depends on the amount of sound absorption in the room. For most practical purposes it may be taken to be equal to the total sound absorption, A_t. The total sound absorption, A_t, may itself be related via the Sabine formula to the reverberation time, and so the formula for the reverberant sound may take alternative forms:

$$L_{rev} = 10\log_{10} W + 6 - 10\log_{10} A_t \quad dB$$

$$L_{rev} = 10\log_{10} W - 10\log_{10} V + 10\log_{10} T + 14 \quad dB$$

It can be seen from the formulae that the reverberant sound level does not vary with distance from the source, and is therefore uniform throughout the room. It does, however, vary with the amount of absorption in the room. A listener hears the direct and reverberant sounds simultaneously, and so L_{dir} and L_{rev} must be combined. This is shown in Fig. 2.5 for three different levels of absorption, in which the sloping line represents the drop in the direct sound as one moves away from the source, and the horizontal lines represent the reverberant sound. The figure also shows that near the source, the overall combined sound level closely follows the direct sound, whereas beyond a certain distance it follows the reverberant sound. Thus the sound field in a room exhibits two distinct regions, the 'direct (or near) field' where the direct sound is dominant, and the 'reverberant (or far) field' in which the reverberant sound is dominant. When the total sound absorption in the room is low, most of the room is in the reverberant field and experiences reverberant sound. Such a room has a long reverberation time and is said to be acoustically 'live'. Conversely, if the sound absorption is high, most of the room is in the direct field, the reverberation time is short, and the room is said to be 'dead'.

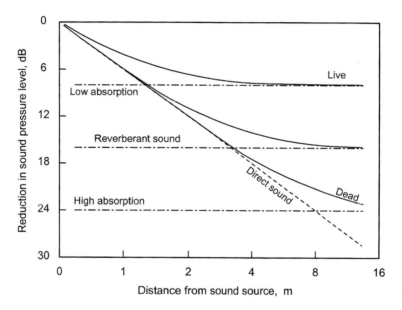

Fig. 2.5   Decay in sound pressure level in a room due to a uniform point source.

In the case of sound transmitted through a separating wall or floor, the resulting sound level in the receiving room is also a combination of a direct field and a reverberant field. However, in this case the sound source is potentially the whole surface of the separating element, which is in essence a planar source. For an infinitely large uniformly radiating planar source, the direct sound would be independent of distance from the plane. The surface of a separating wall is neither infinite nor uniformly radiating, but nevertheless the decrease with distance is much less than for a uniform point source, so that the direct sound field is much more even throughout the receiving room. This fact makes it easier to find microphone positions that will measure the average sound pressure level in a receiving room when conducting sound transmission tests.

2.5   Controlling noise levels by means of absorbing materials

If noise is generated within a space, and there is no possibility of eliminating, reducing or even enclosing the noise source, the options for reducing noise nuisance are limited. This applies to the common areas of multi-occupancy buildings, where the normal noises of occupants entering and leaving cannot be avoided. The nuisance caused is mainly due to the fact that the surface finishes in common areas are often non-porous, with low sound absorption coefficients. This has two effects:

- The reverberation time is long, causing nuisance due to the resulting echoing of noise around the space
- The low sound absorption causes the reverberant sound field to be significant, leading to nuisance due to high sound pressure levels throughout the space.

The inclusion of sound absorbing finishes in the space can ameliorate both problems.

Consider first the reverberation time. To reduce the echo effect, the reverberation time must be kept below a maximum value, say T_{max}. Ignoring the absorption effects of people, furnishings and the air, the Sabine equation can be written to give the total sound absorption of the space in terms of T_{max}:

$$\text{Total sound absorption} \quad \sum_{i=1}^{n} \alpha S = \frac{0.16V}{T_{max}}$$

The surfaces available for sound absorption treatment are usually less than the total, and so the total sound absorption must be written as:

$$\text{Total sound absorption} \quad \sum_{i=1}^{n} \alpha S = \sum_{i=1}^{p} \alpha_{av} S_{av} + \sum_{i=1}^{q} \alpha_{un} S_{un}$$

where α_{av} and S_{av} refer to a total of p surfaces available for treatment
α_{un} and S_{un} refer to the remaining q surfaces not available for treatment.

From this, an expression for the required minimum sound absorption of the *treated* surfaces can be found:

$$\text{Total sound absorption of treated surfaces} \quad \sum_{i=1}^{p} \alpha_{av} S_{av} \geq \left(\frac{0.16V}{T_{max}} - \sum_{i=1}^{q} \alpha_{un} S_{un} \right)$$

For the reduction in reverberation time to be effective, it must be carried out over the whole of the relevant frequency range. Sound absorption data is usually available in octave rather than third octave bands, and according to AD E the calculation should be done at each octave band in the range 250Hz to 4000Hz inclusive.

Although this approach to controlling reverberation is fundamentally correct, the method of compliance recommended in the approved document specifies a minimum amount of sound absorption rather than a maximum reverberation time. The reason for this is that the common spaces of multi-occupancy buildings, especially stairwells, often have a complex geometry, making it difficult to make reliable measurements of the reverberation time. This in turn makes it difficult to check compliance against a specified standard value. This is avoided in the approved document by the specification of a minimum amount of sound absorption. This alternative approach is possible because, according to the above equation, for a fixed value of T_{max} and assuming the sound absorption of the untreated areas is small, the required total sound absorption of the treated areas is proportional to V, the volume of the space. A further simplification is possible if it is assumed that the floor to ceiling height of the space is always at or near a standard value of, say, 2.4 m. The required sound absorption is then proportional to the floor area. It is therefore possible, without significant loss of accuracy, to specify a minimum amount of sound absorption per unit volume or per unit floor area. It is then relatively easy to check that the required area of material with a specified sound absorption coefficient has been provided.

Consider now the reduction in noise levels. Additional sound absorption in a space reduces the reverberant sound field only, and has no effect on the direct sound. Nevertheless there is a reduction in the overall sound pressure level. The effect can be understood from Fig. 2.5. Although this graph was constructed for a rectangular room of conventional shape, it is still broadly applicable to the more complex shapes of common areas, and it is also relevant because the noise sources (voices, footsteps, etc.) are all essentially point sources. It can be seen that at a distance of about 3 m from the source, the addition of absorption can reduce the overall sound pressure level by a significant amount, and at greater distances the reduction is correspondingly increased. The calculation of the magnitude of the reduction is difficult not only because of the shape of a common area, but also because some of the other parameters may not be known, for example the distance from noise source to listener. In general, it is assumed that if the reverberation time is sufficiently low, the noise nuisance due to sound pressure levels will also be low.

A practical problem arises in the choice of sound absorbing materials. This occurs because materials with a high sound absorption coefficient often do not have a hard-wearing surface, and so are susceptible to becoming dirty or damaged, either accidentally or through vandalism. Nor can these surfaces be easily cleaned or re-decorated without affecting their sound absorption properties. For example, applying paint to a sound absorber with a perforated surface may radically alter its absorption coefficient. The maintenance of sound absorbing finishes must therefore be taken into account. Positioning them out of normal reach on ceilings or on walls at a reasonable height above floor level is often the best solution.

3 The Principles of Airborne Sound Insulation

3.1 Sound reduction indices

Figure 3.1 shows a sound wave striking a panel, and then being transmitted through the panel to the other side.

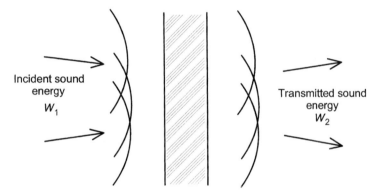

Incident sound
energy
W_1

Transmitted sound
energy
W_2

Fig. 3.1 Sound transmission through a panel.

The sound insulation performance of the panel can be characterised in several ways. The most fundamental measure is the sound transmission coefficient, τ, defined by the equation:

$$\tau = \frac{W_2}{W_1}$$

where:
   W_1 is the sound power incident on the panel
   W_2 is the sound power transmitted by the panel.

In the majority of applications it is more convenient to convert τ to the sound reduction index, R, which is defined by the equation:

$$R = 10 \log_{10}\left(\frac{1}{\tau}\right) = 10 \log_{10}\left(\frac{W_1}{W_2}\right) \quad \text{dB}$$

Figure 3.2 shows a pair of rooms on either side of a panel. Sound is generated in one room (the source room) and is transmitted through the panel, creating sound in the other room (the receiving room). If the sound fields in both rooms are diffuse, and in the absence of flanking transmission, the sound pressure level in the receiving room can be related to the sound pressure level in the source room and the sound reduction index of the panel by:

$$L_2 = L_1 - R + C \quad \text{dB}$$

where L_1 is the sound pressure level in the source room

           L_2 is the sound pressure level in the receiving room

           C is a correction term.

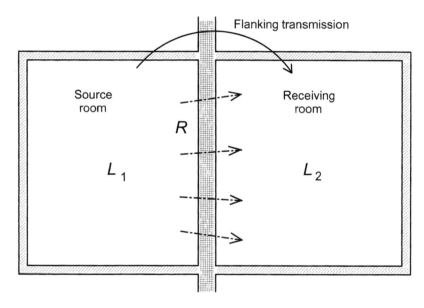

Fig. 3.2   Sound transmission through a separating panel.

The constant C makes allowance for the surface area of the separating panel, and it also includes a correction for reflected sound in the receiving room. The latter is required because the sound pressure level, L_2, in the receiving room consists of the direct sound passing through the panel, plus an additional contribution due to reflection of the direct sound from the receiving room walls. As this reflected sound depends on the acoustic properties of the receiving room and not on the sound transmission properties of the panel, the additional contribution to L_2 may be expressed in terms of the sound absorption of the receiving room, or, by using the Sabine equation, in terms of the reverberation time. The constant C may therefore be written as:

$$C = 10 \log_{10} \left(\frac{S}{A} \right) = 10 \log_{10} \left(\frac{ST}{0.16V} \right)$$

where S is the area of the separating panel
         A is the total sound absorption of the receiving room
         T is the reverberation time of the receiving room
         V is the volume of the receiving room.

If the sound pressure levels in both rooms are measured in order to obtain R, the equation can be rearranged as:

$$R = L_1 - L_2 + 10 \log_{10} \left(\frac{S}{A} \right) = L_1 - L_2 + 10 \log_{10} \left(\frac{ST}{0.16V} \right) \quad \text{dB}$$

An important feature of the sound transmission between two rooms is the presence of flanking transmission. This is sound which finds its way from the source room to the receiving room by any route other than through the separating panel. Specially designed test facilities for measuring R are designed to minimise flanking transmission, but if the measurement is a field test, carried out in a building such as a dwelling, flanking transmission will inevitably influence the result. Because it is a measure of the overall performance of the panel and all its associated structure rather than a measure of the true R value of the panel itself, the result of a field test is often called the apparent sound reduction index, and is defined as:

$$R' = 10 \log_{10} \left(\frac{W_1}{W_2 + W_3} \right) \quad \text{dB}$$

where:
    R' is the apparent sound reduction index of a panel in a building
    W_1 is the sound power incident on the panel
    W_2 is the sound power transmitted through the panel
    W_3 is the sound power transmitted by or through flanking panels and/or other components.

The addition of flanking paths to the sound transmission means that R' is always less than R:

$$R' < R$$

In practice, in a field test it is the overall performance that is required, as the occupants of the receiving room are affected by the total amount of noise coming from the source room, regardless of the route it has taken. It is therefore R' that must be measured.

3.2   Sound reduction through a panel made up of several separate elements

A panel may consist of more than one constructional element, such as a door or window in a wall, as shown in Fig. 3.3.

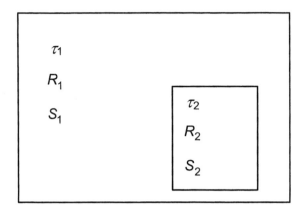

Fig. 3.3   A panel consisting of more than one element.

The average sound transmission coefficient of the whole panel, τ_{av}, may be found with adequate accuracy from the area-weighted coefficients of the component elements:

$$\tau_{av} = \frac{\tau_1 S_1 + \tau_2 S_2 + \tau_3 S_3 + \ldots}{S_1 + S_2 + S_3 + \ldots}$$

where $\tau_1\ \tau_2$ etc. are the transmission coefficients of each element
$\quad\quad$ $S_1\ S_2$ etc. are the areas of each element.

If the sound reduction indices of the elements are known, then their corresponding transmission coefficients can be found by inverting the formula for R:

$$\frac{1}{\tau_1} = \text{antilog}_{10}\left(\frac{R_1}{10}\right)$$

The average sound reduction index, R_{av} is then:

$$R_{av} = 10\log_{10}\left(\frac{1}{\tau_{av}}\right) \quad \text{dB}$$

The most significant consequence of this is that if one element in a panel has a low sound reduction index, then the average for the whole panel will be low. Figure 3.4 is a chart for calculating the change in sound reduction index when a panel consists of two elements.

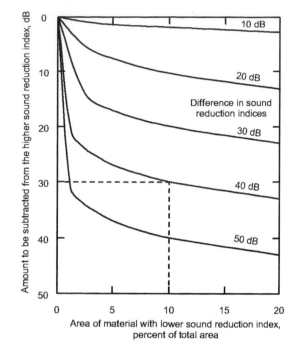

Fig. 3.4   Sound reduction index for a two element panel.

For example, at a particular frequency, suppose that a wall separating two rooms has a sound reduction index of 55dB, and a door connecting the two rooms has a sound reduction index of 15dB when closed. The area of the door, including frame, is 10% of the total wall area. The difference in the sound reduction indices is 40dB, and so, reading the 40dB curve at 10% gives 30dB on the left-hand scale. The effective sound reduction index of the panel is therefore $55 - 30 = 25$dB. The result is closer to the element with the lower sound reduction index, even though it has the smaller area. This is a general principle, and shows that if a panel has within it a relatively small area of poor sound insulation, caused perhaps by faulty workmanship, then that panel will have a substantially reduced acoustic insulation performance.

3.3   The theoretical sound reduction index of a single panel

When a sound wave strikes a panel, the panel is set into a complex pattern of vibrations, the nature of which determines the amount of sound energy transmitted to the room on the other side. The principal factors involved are:

- The frequency of the sound wave
- The mass per unit area of the panel

- The dynamic stiffness (i.e. the elastic property) of the panel
- The damping factor (i.e. the ability to absorb vibrational energy) of the panel.

For a sound wave incident at normal incidence, the sound transmission coefficient is given by:

$$\tau_0 = \frac{(2\,\rho c)^2}{(r + 2\,\rho c)^2 + \left(\omega m - \dfrac{k}{\omega}\right)^2}$$

where τ_0 is the sound transmission coefficient at normal incidence
       ω is the angular frequency of the sound wave ($\omega = 2\pi f$)
       ρ is the density of air
       c is the velocity of sound in air
       r is the damping factor of the panel
       m is the mass per unit area of the panel
       k is the dynamic stiffness of the panel.

This equation can be used to identify three distinct features of the sound transmission coefficient, as the frequency of the sound wave is increased:

(1) At low frequencies, $\dfrac{k}{\omega} \gg \omega m$, and the equation approximates to:

$$\tau_0 \cong \left(\frac{2\,\rho c\omega}{k}\right)^2$$

This shows that the transmission coefficient increases with the square of the frequency of the sound wave. Therefore, the sound reduction index decreases with frequency, and is governed by the dynamic stiffness of the panel:

$$R_0 = 10\log_{10}\left(\frac{1}{\tau_0}\right) = 10\log_{10}\left(\frac{k}{2\,\rho c\omega}\right)^2 = \text{constant} + 20\log_{10}\left(\frac{k}{\omega}\right) \quad \text{dB}$$

This is known as the stiffness controlled frequency region of the sound reduction index.

(2) When the frequency has increased to the point where $\dfrac{k}{\omega} = \omega m$, resonance occurs, and the equation becomes:

$$\tau_0 = \frac{(2\,\rho c)^2}{(r + 2\,\rho c)^2}$$

At this point, only the damping term affects the result, which means that τ_0 is at a maximum, and hence R_0 is at its lowest value. Indeed, if the material has a very low damping coefficient, say $r = 0$, then $\tau_0 = 1$, which means that at this frequency the panel transmits all the sound energy incident upon it and the sound reduction index $R_0 = 0$, i.e. there is no sound reduction at all.

(3)   At high frequencies, $\omega m >> \dfrac{k}{\omega}$, and the equation approximates to:

$$\tau_o \cong \left(\frac{2\rho c}{\omega m}\right)^2$$

This shows that the transmission coefficient decreases with the squares of the frequency of the sound wave and the surface mass of the panel. Therefore, the sound reduction index increases according to the equation:

$$R_0 = 10\log_{10}\left(\frac{1}{\tau_0}\right) = 10\log_{10}\left(\frac{\omega m}{2\rho c}\right)^2 = \text{constant} + 20\log_{10}(\omega m) \quad \text{dB}$$

This is known as the mass controlled frequency region, and the formula indicates that a doubling of either frequency or mass causes R to increase by 6dB.

There are other factors that can have an effect on the sound reduction index of a panel, the most significant being:

- Panel resonances
- The coincidence effect.

The panel resonances occur because the panel is essentially a rectangular plate. In these circumstances, it can be assumed that the plate is held but not clamped at its edges and so the resonant frequencies form a series given by:

$$f_{i,j} \cong 0.45 v_{\text{L}} h \left[\left(\frac{i}{x}\right)^2 + \left(\frac{j}{y}\right)^2\right]$$

where v_{L} is the velocity of longitudinal waves in the material of the panel
h is the thickness of the panel
x, y are the edge dimensions of the panel
i, j are any pair of non-negative integers.

Although there are an infinite number of resonant frequencies, only those at the lowest frequency, when i and j are small, are of significance. Their effect is to cause the sound reduction index to oscillate as the frequency is increased. The velocity of longitudinal waves may be found from:

$$v_{\text{L}} = \sqrt{\frac{E}{\rho(1 - \mu^2)}}$$

where E is the Young's modulus of the material
   ρ is the density of the material
   μ is the Poisson's ratio of the material.

The coincidence effect is caused by bending waves in the panel. Above a certain frequency, known as the critical frequency, the wavelength of the bending waves can coincide with the wavelength of the incident sound wave. This is similar to a resonance effect, and causes an increase in the transmission of sound energy through the panel, i.e. a drop in the sound reduction index below that predicted by the mass law. The critical frequency f_c may be calculated from:

$$f_c = \frac{c^2}{1.8 h v_L}$$

where c is the velocity of sound in air
   h is the thickness of the panel
   v_L is the velocity of longitudinal waves in the material of the panel.

As the frequency of the incident sound wave increases, the drop in R due to coincidence begins at f_c. For diffuse sound fields, the effect becomes most pronounced at approximately $2f_c$, and then gradually fades as the mass law reasserts itself.
   By combining all the factors that affect the sound transmission, it is possible to sketch a theoretical graph of sound reduction index against frequency, as shown in Fig. 3.5.
   Starting at the low frequency end and moving up the frequency scale, the graph shows the characteristics listed overleaf:

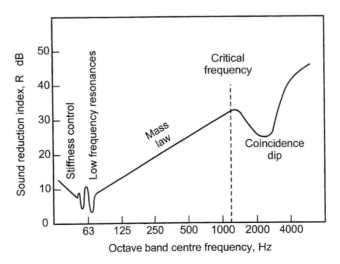

Fig. 3.5   Theoretical variation of sound reduction index with frequency for a single panel.

- At first, at very low frequencies, R reduces as frequency increases. This is the stiffness controlled region.
- R then reaches a minimum at the resonant frequency between the stiffness controlled and mass law regions.
- Next, R oscillates as it passes through the low frequency plate resonances. These often occur at or close to the stiffness/mass resonant frequency.
- Then R increases steadily at a theoretical maximum rate of 6dB per octave (i.e for each doubling of the frequency). This is the mass law region.
- Above the critical frequency, R falls and passes through a minimum, before increasing again with frequency. This is known as the coincidence dip.
- Well above the critical frequency, R approaches the values predicted by the mass law.

3.4   Maximising the sound reduction index of single panels

According to the mass law, the sound reduction index of a single panel of a single homogenous material will increase by 6dB for each doubling of the surface mass. In practice it is found that the increase is less, at about 5dB for each doubling, but even so, increasing the mass of a single panel is the most effective way of raising its overall sound reduction index. For most practical panels, the initial drop in R at low frequency and the low frequency resonances are close to the lower limit of the audible spectrum, and are therefore not important. On the other hand, the dip above the critical frequency can occur anywhere within the audio range, depending on the physical properties and the thickness of the panel. The coincidence effect is therefore of considerable importance. The approximate critical frequencies of some typical materials are given in Table 3.1.

Table 3.1   Approximate critical frequencies for some common materials.

Material	Thickness mm	Critical frequency Hz	Coincidence dip Hz
Reinforced concrete	100	190	380
Reinforced concrete	200	95	190
Brickwork	100	220	440
Brickwork	215	102	204
Plasterboard	26	1500	3000
Plywood	12.5	2500	5000
Plasterboard	13	3000	6000
Glass	5	3000	6000

The critical frequency is inversely proportional to the thickness of a material, and therefore an increase in the thickness pushes the coincidence dip to lower frequencies. It follows that increasing the thickness in order to raise the surface mass of a panel, and hence to improve the sound reduction index by means of the mass law, will also move the coincidence dip to a lower frequency. Thus, although an increase in surface mass will improve the sound reduction index of a single panel at most frequencies, there may be little improvement (or even a reduction) at frequencies near to the new position of the coincidence dip. This is illustrated in Fig. 3.6.

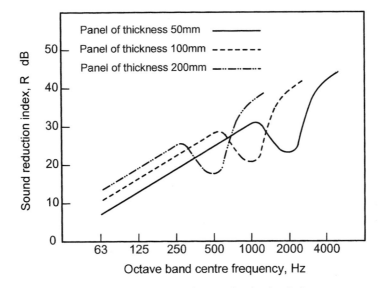

Fig. 3.6   Effect of increasing surface mass on the sound reduction index.

As a result of this phenomenon, the net effect of an increase in surface mass may be much less (or even worse) than that predicted from the mass law alone, because, for dwellings, good sound reduction at low frequency is particularly important.

3.5   Reducing the coincidence effect in single panels

The coincidence effect in single panels can be altered by increasing the stiffness of a panel without increasing its mass. This moves the coincidence effect to higher frequencies, where in certain types of panel the coincidence dip is less damaging. In practice, this can be achieved with 'sandwich' panels, consisting of strong surface layers bonded together by a lightweight non-rigid core, as shown in Fig. 3.7a.

The critical frequency, f_c, of an ideal sandwich panel of this type is given by:

$$f_c = K \left(\frac{m}{t} \right)^{\frac{1}{2}}$$

where K is a constant

m is the mass per unit area of the whole panel

t is the thickness of the stiff surface layer.

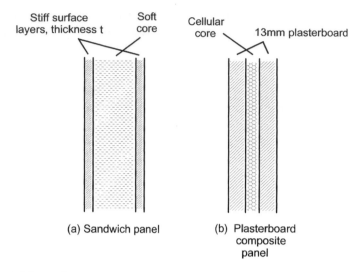

(a) Sandwich panel

(b) Plasterboard composite panel

Fig. 3.7 Sandwich panel.

It can be seen that if the thickness of the surface layer is sufficiently small, then the critical frequency may become too high to be of concern. However, the high bending stiffness of a sandwich panel means that low frequency resonances may become troublesome. The plasterboard composite panel shown in Fig. 3.7b is a typical example.

3.6   The theoretical sound reduction index of a double panel

Double panels consist of two layers of material separated by an air space, examples being a double glazing unit, plasterboard on both sides of timber studs, and a masonry cavity wall. Although many of the characteristics are similar, double panel theory is more complex, and resonances become more important. Starting at low frequency and moving up the frequency scale, the significant features are:

- Low frequency panel resonances (same as for single panels).
- Mass–air–mass resonance. This is caused by the air layer between the panels acting as a spring. The lowest and most important frequency, f_0, at which this occurs is:

$$f_0 = \frac{c}{\pi \cos \theta} \left(\frac{\rho}{(m_1 + m_2)h} \right)^{\frac{1}{2}}$$

where ρ is the density of air

        c is the velocity of sound in air

        θ is the angle of incidence of the sound wave

        m_1, m_2 are the surface masses of the two layers of the panel

        h is the thickness of the air space in the panel.

- Transverse cavity resonances. These are resonances of the air in the air space in the direction parallel to the layers of the panel. Their frequencies are given by:

$$f_{i,j} = \frac{c}{2}\left[\left(\frac{i}{x}\right)^2 + \left(\frac{j}{y}\right)^2\right]^{\frac{1}{2}}$$

where c is the velocity of sound in air

x, y are the edge dimensions of the panels

i, j are any pair of non-negative integers.

- Mass law region. In the case of a double panel, the sound reduction index increases at a maximum theoretical rate of 12dB per octave.
- Coincidence effects. These are the same as for a single panel. If the two layers are of identical material and the same thickness, their critical frequencies will be the same, and the coincidence dip of the double panel will be exaggerated in comparison with that for a single panel. However, if the two layers differ in any respect, their critical frequencies will also differ, and the coincidence effect will be smoothed out. An example of this occurs in high performance sound insulation windows, where the glazing unit is made up of two or more glass sheets of different thickness.
- Longitudinal cavity resonances. These are resonances of the air across the air space in the direction perpendicular to the layers of the panel. Their frequencies are given by:

$$f_n = \frac{nc}{2h\cos\theta}$$

where n is a positive integer

c is the velocity of sound in air

h is the thickness of the air space in the panel

θ is the angle of incidence of the sound wave.

3.7 Maximising the sound reduction index of double panels

Because of the mass law, the natural tendency for the sound reduction performance of a double panel is the same as for a single panel:

- To be poor at low frequency, but to improve as the frequency is increased
- To improve as the surface mass is increased, i.e. as the thickness is increased.

Low frequency resonances and the coincidence effect also apply to double as well as single panels, but the other three resonant modes are characteristics of double panels.

The deleterious effects of the transverse and longitudinal resonances in the air space can be reduced by including sound *absorbing* material in the cavity. This is usually in the form of a fibrous quilt, and by absorbing sound energy reduces the standing wave patterns. The net result is a small but significant increase in the sound reduction index, mostly towards higher frequencies.

The mass–air–mass resonance is the consequence of the air in the cavity behaving like an elastic material that couples the two layers together as though they were connected by a spring. Air at atmospheric pressure is highly compressible and so is a very weak spring, and the effect on the sound reduction index is not large. However, for structural reasons, many double panels consist of two layers of material connected by a mechanical fixing. A particular example is the masonry cavity wall with wall ties. In this type of construction, the mass–air–mass resonance becomes a specific mass–spring–mass system, in which the elastic properties of the wall ties are crucial in determining the sound reduction performance of the panel.

3.8 The effect of wall ties in a double panel

When the two layers of a double panel are mechanically connected, as in a masonry cavity wall with wall ties, the complete structure behaves as a mass–spring–mass system, as shown in Fig. 3.8.

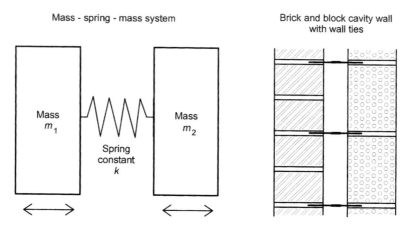

Fig. 3.8 The mass-spring–mass system of a cavity wall with wall ties.

A system consisting of a single body vibrating against a solid immoveable surface has a natural resonant frequency, f_0, which is given by the equation:

$$f_0 = \frac{1}{2\pi}\sqrt{\frac{k}{m}}$$

where k is the spring constant, or dynamic stiffness
            m is the mass of the vibrating body.

For a fuller description of the underlying vibration theory and a definition of dynamic stiffness, see Chapter 4, section 4.5.

In the case of a cavity wall, there are two approximately similar masses vibrating together, and the formula for the resonant frequency becomes:

$$f_0 = \frac{1}{2\pi} \sqrt{\frac{nk}{M}}$$

where n is the number of wall ties per unit area of the wall
            k is the dynamic stiffness of one wall tie
            M is a combination of the surface masses of the two layers

$$M = \frac{m_1 m_2}{m_1 + m_2}$$

where m_1, m_2 are the masses per unit area of layers 1 and 2 respectively.

For good acoustic isolation between the two layers it is desirable for the resonant frequency to be as low as possible, preferably below 50 Hz. By substituting $f_0 = 50$ into the equation, the maximum permissible dynamic stiffness of the wall ties can be calculated. Typical values of the dynamic stiffness range from about 1.7 MN/m for butterfly ties to over 80 MN/m for vertical twist ties. From this it can be concluded that, in order to achieve a high sound reduction index in a typical masonry cavity separating wall, if wall ties are necessary the butterfly type is the most appropriate.

3.9   Other indices of sound reduction between rooms

Because the result of a field test yields the sound reduction performance of the complete structure rather than the panel itself, a number of alternative measurements of sound reduction have been defined and are in common use. For the purposes of AD E, the important ones are as follows.

The normalised level difference, D_n (as defined in ISO 140-4:1998)

$$D_n = L_1 - L_2 - 10 \log_{10}\left(\frac{A}{A_0}\right) \quad \text{dB}$$

where:
            D_n is the normalised level difference
            L_1 is the average sound pressure level in the source room
            L_2 is the average sound pressure level in the receiving room
            A is the equivalent sound absorption area of the receiving room

A_0 is a reference absorption area (for rooms in dwellings or rooms of comparable size, $A_0 = 10\,\text{m}^2$).

The standardised level difference, D_{nT} (as defined in ISO 140-4:1998)

$$D_{nT} = L_1 - L_2 + 10\log_{10}\left(\frac{T}{T_0}\right) \quad \text{dB}$$

where:

D_{nT} is the standardised level difference
L_1 is the average sound pressure level in the source room
L_2 is the average sound pressure level in the receiving room
T is the reverberation time in the receiving room
T_0 is a reference reverberation time (for rooms in dwellings, $T_0 = 0.5\,\text{s}$).

3.10   Conversion of sound reduction indices to single number quantities

The above airborne sound insulation properties are normally measured in third octave bands over a representative part of the audible spectrum, usually from 100Hz to 3150Hz. This gives 16 separate values of R' (or D_n etc.), one for each third octave band. It would be a simple matter to calculate the arithmetic average of these 16 R' values, and then to take this average as an indicator of the overall performance of the panel. However, this is unsatisfactory for two reasons. Firstly, a simple average takes no account of the increase in the sensitivity of the ear with frequency. A very low value of R' in a high frequency band may be more significant in its effect than the same low value in a low frequency band. Secondly, a very low value in any one of the frequency bands becomes hidden in the average by a correspondingly high value in another. As far as a listener is concerned, the effectiveness of the panel in providing sound insulation will be determined by the very low value rather than the average. Thus, for a panel to be deemed satisfactory for sound insulation purposes it is necessary to examine R' in all 16 of the third octave frequency bands. This can be cumbersome and time-consuming, and makes it difficult to make quick comparisons between different sets of results, or between a set of results and a specified standard. Consequently, it is useful to be able to convert the 16 individual third octave band values into a single number index that can then be used as an indicator of performance. There are several ways of doing this. The method required by Part E of the building regulations is that specified in BS EN ISO 717-1:1997. In this, the measured results are compared with a set of reference values, as shown in Table 3.2, over the same range of third octave bands as the measurements.

As the frequency increases from 100Hz to 1250Hz, so also the reference values increase. This has the effect of applying a weighting in favour of the higher frequencies, and so allows for the fact that the ear is more sensitive to medium and

Table 3.2 Third octave band reference values.

Third octave band centre frequency, Hz	Reference value, dB
100	33
125	36
160	39
200	42
250	45
315	48
400	51
500	52
630	53
800	54
1000	55
1250	56
1600	56
2000	56
2500	56
3150	56

high frequencies than low ones, and thus takes account of the first of the objections to calculating a simple average. These reference values are then plotted on the same graph as the measured results for a panel, and the two sets of values are compared at each frequency. If the reference value is the higher, the difference $R_{ref} - R'$ is noted, otherwise it is ignored, and the sum of these differences (known as the adverse deviation) is found. The reference curve is then moved up or down, in steps of 1dB at a time, and the adverse deviation recalculated until a position is found where the adverse deviation is as large as possible, but not more than 32.0dB. When this procedure is complete, the decibel value of the reference curve at 500Hz is read, and this is taken as the 'Single number weighted apparent sound reduction index' of the panel. The same procedure and the same set of reference values may be used to convert any other third octave band measurement of sound insulation to a single number, the important ones being shown in Table 3.3.

Unfortunately, although better than taking a simple arithmetic average of the 16 third octave bands, the single number weighted indices do not fully take account of the effect of a very low value of R' in one of the frequency bands. This is of most importance at low frequencies, where the sound insulation of many separating structures is weakest, and where the noise nuisance due to powerful low frequency sources is greatest. This problem is addressed by means of the spectrum adaptation term.

3.11   The spectrum adaptation term

Because the reference values in Table 3.2 reduce as the frequency is reduced, the calculation of the weighted single number indices automatically makes some allowance for:

Table 3.3   Third octave band values and weighted single number equivalents.

Third octave band measurement		Weighted single number equivalent	
Sound reduction index	R	Weighted sound reduction index	R_w
Apparent sound reduction index	R'	Weighted apparent sound reduction index	R'_w
Normalised level difference	D_n	Weighted normalised level difference	$D_{n,w}$
Standardised level difference	D_{nT}	Weighted standardised level diffeence	$D_{nT,w}$
Standardised level difference	$D_{ls,2m,nT}$	Weighted standardised level difference	$D_{ls,2m,nT,w}$
	$D_{tr,2m,nT}$		$D_{tr,2m,nT,w}$

- The variation in the sensitivity of the human ear with frequency
- The fact that sound reduction indices tend to be less at low frequencies because of the mass law.

However, the noise nuisance suffered by a listener due to sound passing through a panel will depend not only on the weighted single number index of the panel (R, R', D_n or D_{nT}, whichever is the most appropriate for the circumstances) but also on the frequency characteristics of the source of noise itself. Two panels with same value of D_n may appear to provide different levels of sound reduction if the noise source has a frequency spectrum with a bias, say, for example, towards low frequencies. Furthermore, the relative performance of the panels may change (i.e. the poorer may become the better) if the noise source is different, say with a bias towards high frequencies. The purpose of the spectrum adaptation term is to adjust the value of the weighted single number index to take account of the frequency spectrum of the type of noise for which the panel must provide sound insulation. In principle, it is possible to create a spectrum adaptation term for a noise with any specified frequency characteristic, but the noises of particular concern to the occupants of buildings are:

- Domestic noise
- Traffic noise
- Aircraft noise.

There is some overlap in the characteristics of these noises. Two frequency spectra are defined by **BS EN ISO 717-1**, and their third octave values for the range 100 to 3150Hz are shown in Table 3.4. It should be noted that all the decibel levels in this table are the values that would be measured using the A-weighted network. The first spectrum, C, takes account of noise from:

- Domestic activities, including speech, music, radio and television
- Children playing
- Railway traffic at medium and high speed
- Road traffic at speeds greater than 80 km/hour
- Jet aircraft at short distance.

The second spectrum, C_{tr}, takes account of noise from:

- Urban road traffic
- Railway traffic at low speeds
- Propeller driven aircraft
- Jet aircraft at large distance
- Disco music
- Factories emitting mainly low and medium frequency noise.

Table 3.4   Sound spectra for calculating adaptation terms.

Frequency Hz	Spectrum No. 1 C, dB	Spectrum No. 2 C_{tr}, dB
100	−29	−20
125	−26	−20
160	−23	−18
200	−21	−16
250	−19	−15
315	−17	−14
400	−15	−13
500	−13	−12
630	−12	−11
800	−11	−9
1000	−10	−8
1250	−9	−9
1600	−9	−10
2000	−9	−11
2500	−9	−13
3150	−9	−15

Only the shape of these spectra is important, and so the values in the frequency bands are adjusted (i.e. normalised) so that overall spectrum has a sound pressure level of 0dBA.

The approved document states that the adaptation term C_{tr} is to be used to correct measured values of weighted sound reduction indices. The reasons for C_{tr} being preferred to C are:

- Many of the complaints of noise intrusion into dwellings via walls cite low frequency sound, especially from powerful music systems or from the rumble of urban transport

- The C_{tr} spectrum is biased towards noises with a significant low frequency content.

The net effect of using C_{tr} can therefore be expected to be an increase in the weighted single number index of a panel that has a good sound insulation performance at low frequency. Conversely, a panel with a poor low frequency performance but the same weighted single number index (due to better performance at high frequency) is likely to have its index reduced by the application of C_{tr}.

3.12 Reducing airborne sound transmission – implications for building design

In order to reduce the airborne sound transmission between adjacent rooms it is necessary to pay attention to three sets of parameters:

- The design of the separating panel itself
- Flanking transmission via paths through adjoining components
- The quality of workmanship in the construction of the panel and all its associated structure.

For single panels, the general rule, derived from the mass law, is that an increase in the mass per unit area of the panel will increase its sound reduction index. This, however, is subject to the effects of panel resonances and the coincidence effect. For panels of two or more layers the mass per unit area is still important, but the main principle is to minimise or even eliminate mechanical connection between at least one pair of layers. Resonance effects in multiple panels are more complex, and must also be taken into account. Calculation of the approximate sound reduction index of a panel from first principles may sometimes be possible, but is neither accurate nor reliable enough when trying to meet a specific target such as a regulatory standard. In any case calculation from first principles is unnecessary, as there is an extensive literature giving design details of separating panels for achieving good sound reduction. The approved document itself contains guidance on suitable designs, and there are several other sources, for example, [9], [11], [12], as well as data in trade literature.

Flanking transmission occurs in at least four ways:

- Via vibration paths created by the structural connections of the separating panel to the rest of the building structure
- Via vibration paths between all the other elements of the source room (floor, ceiling, other walls) and the rest of the building structure
- Via airborne sound transmission through a side wall (or floor or ceiling) into another room (or space), and then from that room into the receiving room
- Via air paths created by services installations, such as air ducts for ventilation or warm air heating systems, chimneys and flues.

The first two of these are especially difficult to control, because the degree of flanking transmission depends heavily not only on the exact detailing of the edge connections of the panel but also on the surrounding structure to which it is attached. It is therefore necessary to consider the panel and its associated structure as a whole. Calculation of the effects of edge connections is most unlikely to be possible, and so the most satisfactory method of testing a design is by full scale measurement in a completed building. Laboratory measurement is possible in some cases, but only in an acoustic facility which has been specially designed to allow for flanking transmission, and only if all relevant parts of the structure are correctly reproduced in the laboratory arrangement. Of the other possible types of flanking transmission, that due to air paths can have the biggest effect, and so care must be taken, especially in flats and similar residential buildings, to ensure that the sound reduction of a separating panel is not short-circuited by, say, a ventilation or heating system. The air path need not be a direct connection between the rooms on either side of the panel, but between one of those rooms and some other space in the building. The guidance given in AD E on the construction of separating panels includes guidance on the surrounding structure in order to control flanking transmission.

The final performance of a separating panel is critically dependent on precise execution of the design at the construction stage. There are several ways in which the sound reduction performance can be reduced by poor or inappropriate workmanship, for example:

- Incomplete filling of mortar joints in brick or block
- Incomplete sealing of joints around the edges of panels
- Incorrect or careless installation of services, especially electrical sockets, in the separating panel, or unplanned installation of such services
- Forming rigid or incorrect connections between layers of a multiple layer panel, or between a layer and its edge connections.

Many of these will create an air path through part of the structure, which, even if it is a small proportion of the area, can substantially affect the overall sound reduction. Furthermore, workmanship faults are almost always hidden from view in the completed construction. This is one of the reasons for the introduction of pre-completion testing.

4 The Principles of Impact Sound Insulation

4.1 Impact sound reduction indices

Impact noise is generated not by a noise source but by the impact of one object against another. The vibrations generated by the impact travel through the structure by many different routes, creating noise in many parts of the building, but mainly in the room on the other side of the building element that is being struck. As far as Part E of the Building Regulations is concerned, the most important type of impact noise is that caused by objects striking the floor of a building, such as footsteps, the movement of furniture, articles being dropped onto the floor, etc., and so the room most affected by the resulting noise nuisance is the room below.

In order to measure the effect of impact sound on a floor/ceiling structure it is conceptually possible to begin by taking the difference in the sound pressure levels between the two rooms. However this is unsatisfactory, because the sound pressure level created in the source room depends on the properties of the source room as well as on the type and intensity of the impact. The preferred approach is to specify a standard method of generating impacts, and to measure the sound pressure level in the receiving room only. The sound pressure level in the receiving room, suitably weighted, is then used directly as the criterion of performance. Clearly, the lower the sound level in the receiving room the better is the resistance of the structure to impact noise, and so whereas the criteria for airborne sound specify minimum values for sound reduction, the criteria for impact noise specify maximum values for the noise level that is generated.

Consider a pair of rooms as shown in Fig. 4.1. A standard impact or tapping machine is placed on the floor of the upper room, and is set to produce a continuous series of rapid impacts.

The sound pressure level in the lower (receiving) room is measured in third octave bands over the usual range of 100Hz to 3150Hz. As with the measurement of airborne sound transmission, the sound levels in the receiving room are affected by its sound absorption. In order to make comparisons and to set standards, the measured results are corrected to a room with standardised sound absorption properties. If the pair of rooms is part of an acoustics test facility, and the results derive from a laboratory measurement, the correction is given by:

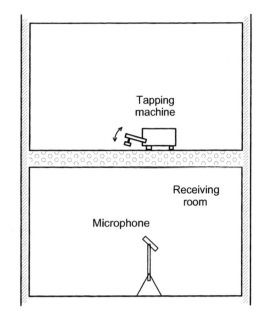

Fig. 4.1 Pair of rooms with tapping machine.

$$L_{\mathrm{n}} = L_{\mathrm{rec}} + 10 \log_{10}\left(\frac{A}{A_0}\right) \quad \mathrm{dB}$$

where L_{n} is the *normalised* impact sound pressure level, laboratory measurement

L_{rec} is the measured sound pressure level

A is the total sound absorption in the receiving room

A_0 is a reference absorption area.

The usual value for the reference absorption is $A_0 = 10\,\mathrm{m}^2$.

If the pair of rooms is in a building, particularly a dwelling, and the results derive from a field measurement, the correction may be expressed either in terms of total sound absorption or as a standard reverberation time. The two methods of correction lead to slightly different measures of impact sound resistance:

$$L'_{\mathrm{n}} = L_{\mathrm{rec}} + 10 \log_{10}\left(\frac{A}{A_0}\right) \quad \mathrm{dB}$$

where L'_{n} is the *normalised* impact sound pressure level, field measurement

L_{rec} is the measured sound pressure level

A is the total sound absorption in the receiving room

A_0 is a reference absorption area.

The usual value for the reference absorption is again $A_0 = 10\,\mathrm{m}^2$. The alternative is:

$$L'_{nT} = L_{rec} - 10\log_{10}\left(\frac{T}{T_0}\right) \quad \mathrm{dB}$$

where L'_{nT} is the *standardised* impact sound pressure level, field measurement
L_{rec} is the measured sound pressure level
T is the reverberation time of the receiving room
T_0 is a reference reverberation time.

For dwellings, the usual value for the reference reverberation time is $T_0 = 0.5\,\mathrm{s}$.

4.2   Conversion of impact sound indices to single number quantities

As with airborne sound insulation, the above impact sound insulation properties are normally measured in third octave bands over the range 100Hz to 3150Hz, giving 16 separate values of L_n, (or L'_n or L'_{nT}), one for each third octave band. Again, it is useful to be able to convert the 16 individual third octave band values into a single number index that can then be used as a performance indicator, and the method of doing this is in principle the same as that used for airborne sound. The measured results are compared with a set of reference values, specified in BS EN ISO 717-2:1996, and shown in Table 4.1 for third octave bands.

It should be noted that the reference values are sound pressure levels in the receiving room, and so as the frequency increases from 400Hz to 3150Hz, so the

Table 4.1   Third octave band reference values.

Third octave band centre frequency, Hz	Sound pressure level reference value, dB
100	62
125	62
160	62
200	62
250	62
315	62
400	61
500	60
630	59
800	58
1000	57
1250	54
1600	51
2000	48
2500	45
3150	42

reference values decrease, effectively weighting the higher frequencies to allow for the greater sensitivity of the ear at high frequencies. These reference values are then plotted on the same graph as the measured results, and the two sets of values are compared at each frequency. If the measured value exceeds the reference value, the difference $L_n - R_{ref}$, (or $L'_n - R_{ref}$ or $L'_{nT} - R_{ref}$) is noted, otherwise it is ignored, and the sum of these differences (known as the adverse deviation) is found. The reference curve is then moved up or down, in steps of 1dB at a time, and the adverse deviation recalculated until a position is found where the adverse deviation is as large as possible, but not more than 32.0dB. When this procedure is complete, the 'Single number weighted impact sound pressure level' is found by subtracting 5dB from the decibel value of the reference curve at 500Hz. The procedure may be applied to all three impact sound measurements, as detailed in Table 4.2.

Table 4.2   Third octave band values and weighted single number equivalents.

Third octave band measurement		Weighted single number equivalent	
Normalised impact sound pressure level	L_n	Weighted normalised impact sound pressure level	$L_{n,w}$
Normalised impact sound pressure level	L'_n	Weighted normalised impact sound pressure level	$L'_{n,w}$
Standardised impact sound pressure level	L'_{nT}	Weighted standardised impact sound pressure level	$L'_{nT,w}$

4.3   The spectrum adaptation term

Spectrum adaptation terms can be defined for impact sound as well as for airborne sound, but in the case of impact sound their purpose is to correct for variations in the type of impact. The normal measurement of impact noise utilising a continuously operating tapping machine produces a result that is appropriate for sources with similar characteristics, such as footsteps. However, the result may not be suitable for single impacts on particular types of floor. To provide for this, BS EN ISO 717-2:1996 defines a spectrum adaptation term, C_I, which may be calculated from any of the three single number quantities as follows:

$$C_I = L_{n,sum} - 15 - L_{n,w} \quad \text{dB}$$
$$C_I = L'_{n,sum} - 15 - L'_{n,w} \quad \text{dB}$$
$$C_I = L'_{nT,sum} - 15 - L'_{nT,w} \quad \text{dB}$$

where $L_{n,w}$, $L'_{n,w}$ and $L'_{nT,w}$ are as defined in Table 4.2

$L_{n,sum}$, $L'_{n,sum}$ and $L'_{nT,sum}$ are summations obtained from the results of the relevant measurements of sound pressure level in the third octave bands from 100Hz to 2500Hz

The $L_{n,sum}$ etc. are the decibel addition (i.e. the addition on an energy basis) of the third octave band values, i.e.:

$$L_{n,sum} = 10 \log_{10} \sum_{i=1}^{k} 10^{L_{n,i}/10} \quad dB$$

where $L_{n,i}$ is the sound pressure level in band number i, and k is the number of bands (15 in this case as the 3150Hz band is not included). Although the spectrum adaptation term for impact sound may be useful in certain cases, it does not appear in AD E.

4.4   Reduction in impact sound due to an added floor covering

It is sometimes necessary to know the effect that a particular floor covering may have in reducing the impact sound pressure level in the room below. This is most often the case with a covering applied to a bare concrete slab. Unfortunately, the performance of a covering varies with the behaviour of the floor slab to which it is applied, and so **BS EN ISO 717-2:1996** defines impact sound pressure levels for a reference floor. When measurements are carried out on this reference floor, and the results processed, the single number weighted normalised impact sound pressure level is expected to be 78dB. Laboratory measurements are then carried out with the covering applied to the reference floor, and the single number weighted normalised impact sound pressure level of the reference floor plus its covering, denoted by $L_{n,r,w}$, is found. Clearly, $L_{n,r,w}$ should be less than 78dB, and so the single number weighted reduction of impact sound level for the floor covering itself, ΔL_w, is found from:

$$\Delta L_w = 78 - L_{n,r,w}$$

In principle, the single number weighted reduction of impact sound level can be measured for any covering applied to the reference floor. If the covering is itself a rigid construction such as a floating floor mounted on a resilient layer, then AD E specifies that two tests should be carried out, one with the floor loaded by a distributed load similar to that expected in normal service, and the other with the floor unloaded.

   In general, it is better to measure the impact sound reduction of a complete floor structure, complete with all allowable floor coverings and with the ceiling beneath. However, in some circumstances, for example when a soft covering is applied to a concrete base, it is necessary to know the ΔL_w of the floor covering itself. Chapter 8 gives details.

4.5   Reducing impact sound transmission – the theory of vibration isolation

The effect of an impact on the floor (or any other building component) is to inject vibrational energy directly into the building structure. The vibrations may be transmitted to any part of the building to which there is a structural link, but most obviously to the room beneath the floor. The ease with which the vibrations are transmitted depends on the elastic properties of the materials through which they must pass, and the connections between them. The exact mechanisms are complex, but in general the more elastic and resilient the materials, the more energy that is absorbed within them, and the less that is transmitted to another room.

The elementary basis of vibration isolation is shown in Fig. 4.2. A body of mass m_1 is attached to another body of mass m_2 via a spring of spring constant (or dynamic stiffness) k, and the bodies vibrate parallel to the direction of the spring. The system has a natural (resonant) frequency of vibration, f_0, determined by the masses and the dynamic stiffness of the spring:

$$f_0 = \frac{1}{2\pi} \sqrt{\frac{k}{M}}$$

where

$$M = \frac{m_1 m_2}{m_1 + m_2}$$

The dynamic stiffness of the spring is analogous to Young's modulus for a static system, and is similarly defined as force per unit displacement:

$$k = \frac{\text{force}}{\text{displacement}} = \frac{F}{\delta x}$$

where F is the amplitude of the dynamic force on the spring
         δx is the amplitude of the dynamic displacement.

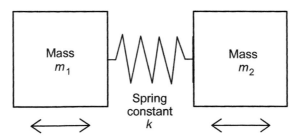

Fig. 4.2   A mass–spring–mass system.

The value of k depends on the length of the spring, i.e. the separation of the two masses, and so it is common practice to indicate the separation as a subscript, $k_{X,mm}$, where X is the static separation of the masses in millimetres. The measurement of the dynamic force and displacement is difficult, and so the dynamic stiffness is often obtained from a measurement of the resonant frequency:

$$k_{X,mm} = 4\pi^2 M f_0^2$$

If the two masses are equal, i.e. $m_1 = m_2 = m_{av}$, then we may write:

$$M = \frac{m_{av}^2}{2m_{av}} = \frac{m_{av}}{2}$$

and the equation for the dynamic stiffness becomes:

$$k_{X,mm} = 2\pi^2 m_{av} f_0^2$$

In the case of a floating floor, the structural or sub-floor is normally of a much higher mass than the floating floor itself. The system can be approximated to a single mass vibrating via a spring against a body of infinite mass. In this case, the combined mass M is equal to the total mass, m_{ff}, of the floating floor alone, i.e:

$$M = m_{ff}$$

In this system, the vibrating mass transfers a force through the spring to the surface to which it is attached. Ignoring the effect of damping, the force transmitted to the surface can be expressed by:

$$F_{tr} = \pm \left(\frac{1}{1 - \left(\frac{f}{f_0} \right)^2} \right)$$

where F_{tr} is the force transmissibility, i.e. the force transmitted to the surface per unit of applied force (if the formula gives a negative result, it is taken as positive)

f is the applied frequency, i.e. the frequency at which the mass is being made to vibrate

f_0 is the natural frequency of the mass/spring system in the absence of damping.

Examination of the equation shows that as the applied frequency is increased from zero, the force transmissibility increases, approaches a maximum, becoming infinity at the point of resonance when $f = f_0$. Above f_0 the transmissibility decreases, eventually approaching zero as f approaches infinity. This is shown in Fig. 4.3. In this, values of F_{tr} greater than one are an amplification of the vibration effect.

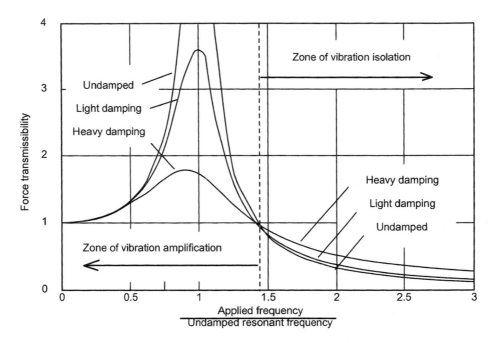

Fig. 4.3   Force transmissibility curves for a one-dimension mass–spring system.

Values less than one are an attenuation, and so indicate a degree of vibration isolation.

For the simple system illustrated, to which Figs 4.2 and 4.3 apply, the transition from vibration amplification to vibration isolation occurs at $f = \sqrt{2}f_0$. In a practical situation, some of the vibrational energy will be dissipated by damping in the material of the spring, and the effect of this is also shown in Fig. 4.3. As the degree of damping is increased, two effects occur. First, the effect of the mass–spring system on the transmitted force becomes less pronounced, with reduced amplification at frequencies below the transition and reduced vibration isolation at frequencies above it. Second, the point of resonance moves to lower frequencies, so that the new resonant frequency, f_r, is below the natural resonant frequency, f_0. Although the vibrations engendered by random impacts on the floors of buildings are considerably more complex than the situation considered here, several general conclusions can be drawn from this elementary theory:

- Vibration isolation occurs only for frequencies well above the natural frequency of the mass–spring system
- The frequency range over which vibration isolation occurs can be increased by making the natural frequency as low as possible
- At low frequencies, amplification of the applied vibration occurs.

These conclusions suggest how the mass–spring system should be designed in terms of its principal properties, namely the spring constant, the mass and the damping.

Firstly, in order to provide vibration isolation over a large frequency range, the natural frequency, f_0, of the mass–spring system should be as low as possible. This is necessary because mechanical impacts generate frequencies over a wide range. The formula for f_0 shows that this can be achieved either by reducing the spring constant or by increasing the mass. However, reducing the spring constant and/or increasing the mass result in an increase in the deflection of the spring under an applied load. For the mass–spring system to be effective, it must have sufficient space to take up this deflection. Clearly, for practical reasons there is a limit to the amount of deflection of a building component that would be considered tolerable, and so a compromise must be struck between the competing requirements of achieving a low resonant frequency while at the same time preventing excessive deflection of any part of the building structure.

Secondly, in order to control the vibration amplification that occurs at low frequency, it is necessary to include damping in the system. However, as can be seen from Fig. 4.3, the greater the degree of damping the lower the effectiveness of vibration isolation at high frequency. Again, a compromise must be struck between competing requirements.

4.6   Reducing impact sound transmission – implications for building design

The intensity of the vibrational energy is greatest close to the point of impact, and so control measures are most effective when placed here. In the case of impacts on the floor of a building, an impact can occur anywhere over the floor area, and the whole floor surface is therefore a potential point of impact. One of the most practical ways of providing a damped spring at the point of impact is to cover the whole floor area with a resilient material. Any resilient material will have some effect. Conventional pile carpeting laid on a foam rubber underlay can be an effective first line of defence against light impacts such as normal footsteps, but unless it is exceptionally thick is unlikely to provide sufficient deflection for heavy impacts. However, any floor covering must be a permanent part of the building fabric in order to be considered as a contributor to the reduction of impact sound, otherwise it cannot be assumed that it will be maintained and replaced when necessary, or even that it will be installed by the occupier. This means that if a soft covering is part of the floor design, that covering must be fixed to the floor in such a way that it can only be removed for the purposes of replacement or refurbishment. Solutions to the transmission of impact noise therefore require consideration of the structure as a whole. The principles to be followed are:

- Provide the sound control measure close to the source
- Provide a spring system with a low natural frequency, either by means of a low spring constant or a high sprung mass, or both
- Ensure that the spring system is not bypassed by direct structural contact either deliberately via, say, a mechanical fixing, or accidentally by contact between components on either side of the spring

- Ensure that there is damping in the system to prevent amplification of low frequency components in the impact.

Practical solutions for meeting these principles include:

- Floating floor construction
- Suspended ceiling beneath the floor
- A structurally heavy floor slab with a permanent resilient floor covering.

In general, because it incorporates a true mass–spring system close to the source of the impacts, the floating floor construction may be expected to be the most effective in reducing nuisance due to impact generated noise. The floating layer may consist of sheet materials such as timber or plywood, or a slab of solid material such as concrete, or any other suitable construction. Similarly, the structural floor upon which the floating layer rests may consist of several possible constructions. The spring that separates the floating and structural floors is normally provided by a layer of resilient material. The contact between the layers may be continuous over the whole area, or at discrete positions where, for example, joists are in alignment. The resilient layer itself may be dimpled or ridged so that contact occurs over a certain fraction of its area. In special applications, such as the floor of a room containing machinery, individually designed mounts could be used. Figure 4.4 illustrates the principal alternative types of floating layer.

An estimate of the natural frequency, f_0, of the floating floor can be obtained from:

$$f_0 = \frac{1}{2\pi}\sqrt{\frac{k}{m}}$$

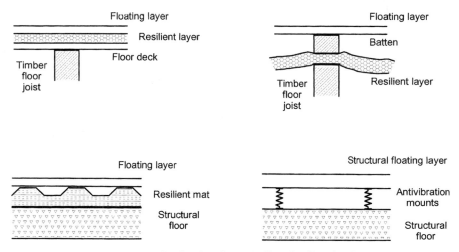

Fig. 4.4 Types of resilient layer for floating floors (ceiling detail omitted).

where k is the dynamic stiffness per unit area of the resilient layer
           m is the mass per unit area of the floating floor.

Ideally, a floating floor should provide vibration isolation over the whole of the relevant frequency range. The lowest third octave band used for measurement is centred at 100Hz, and if this is to be included, the natural frequency must be even lower, at around 50 to 60Hz. Typical materials include:

- Mineral fibre or glass fibre quilts
- Polyethylene or polystyrene foamed materials such as EPS board
- Anti-vibration mat, usually of silicone rubber.

The dynamic stiffness of fibre quilts is less than that of EPS board, making it easier to achieve the required low natural frequency. Anti-vibration mats can be designed to suit a variety of load and isolation requirements. In addition, however, the material must retain its resilient properties over the design life of the floating floor that it supports.

The same theory can be applied to determining the natural frequency of a suspended ceiling, the difference being that the ceiling will be suspended from a resilient fixing, rather than resting on a continuous layer of resilient material. Also, it may be possible to tolerate a larger deflection in a ceiling system than in a floating floor.

Apart from being able to calculate the resonant frequency, it is difficult, if not impossible, to predict from theory the exact impact sound resistance of any construction. This is because the vibrational modes of a building structure, and the connecting paths between the source of the impact noise and the receiving room, are not only complex, but because of differences in workmanship may be unique to each individual building. The most satisfactory method of determining the impact sound resistance of a separating element is by direct measurement in a completed building. In certain circumstances, laboratory measurements may be possible in an acoustic facility that has been specifically designed for flanking transmission.

PART 2
APPROVED DOCUMENT E

Part E of Schedule 1 to the Building Regulations 2000 (as amended) stipulates requirements relating to the transmission of sound within buildings. Approved document E (2003), outlines the performance required of various components of construction in order to comply with the Regulations. The approved document also provides considerable guidance as to how the necessary levels of insulation may be achieved.

The content of Approved Document E (2003) is outlined and explained in Part 2 of this book.

Part E of the Building Regulations and the approved document make frequent reference to 'dwelling houses, flats and rooms for residential purposes'. For convenience, these will, within Part 2 of this book, be referred to collectively as 'dwelling places'. Also, following each section heading within Part 2 there is a list of the sections of the approved document to which it relates. However, since material relating to a specific subject may occur in a number of places within the approved document, these reference lists may not be totally complete or comprehensive.

5 Performance Requirements

Approved Document E Section 0: Performance

5.1 Sound insulation between dwelling places   AD E: 0.1–0.8

Requirement E1 of Schedule 1 to the Building Regulations 2000 (as amended) states that the design and construction of houses, flats and rooms for residential purposes should be such that there is reasonable resistance to sound transmission between them and other parts of the same building and other buildings with which they are in direct physical contact. It should be noted that this requirement applies not only to purpose-built units but also to those resulting from the material change of use from some other type of building. Rooms for residential purposes include rooms in residential homes, student halls of residence and hotels in which people may be expected to sleep.

In circumstances where Requirement E1 applies, Regulation 20A of the Building Regulations 2000 (as amended) and Regulation 12A of the Building (Approved Inspectors etc.) Regulations 2000 (as amended) apply and these regulations impose a requirement to undertake an approved programme of sound insulation testing and provide the appropriate local authority with a copy of the test results. In the case of both regulations there are specific time limits within which the results must be presented to the building control body. It is stated in Approved Document E that the normal way of satisfying regulation 20A or 12A will be to undertake a programme of testing in accordance with section 1 of the approved document. The nature and extent of this pre-completion testing is considered in Chapter 6.

Requirement E1 will be satisfied for separating walls and floors together with stairs which separate dwelling units if they are constructed such that they satisfy the sound insulation values given in Tables 1a and 1b in the Approved Document and also in Table 5.1 below. Separating floors are defined as floors which separate flats or rooms for residential purposes and separating walls are those which separate adjoining dwelling houses, flats or rooms for residential purposes. It is important to note that sound will not only be transmitted directly through the element in question but also through adjacent, or flanking, elements and that when assessing compliance with Table 5.1, flanking as well as direct transmission must be taken into account. See Chapter 3 for explanations of these terms.

Requirement E1 does not relate only to walls and floors which separate dwelling units. If a wall or floor separates a dwelling unit from a space used for communal or non-domestic purposes, e.g. a refuse chute, the level of sound insulation required

Table 5.1 Performance standards for separating walls, separating floors and stairs that have a separating function.

	Minimum value of airborne sound insulation $D_{nT,w} + C_{tr}$ dB	Maximum value of impact sound insulation $L'_{nT,w}$ dB
Purpose built dwelling houses and flats		
Walls (inc. walls that separate houses and flats from rooms for residential purposes)	45	–
Floors and stairs	45	62
Dwelling houses, flats and rooms for residential purposes formed by material change of use		
Walls	43	–
Floors and stairs	43	64
Purpose built rooms for residential purposes		
Walls	43	–
Floors and stairs	45	62

may be greater than that given in Table 5.1 for separating walls and floors. In such circumstances the required value of sound insulation will depend on the level of noise generated in the non-domestic space and expert advice should be sought to assess the anticipated level and, if necessary, recommend higher levels of sound insulation. Figure 5.1 summarises the obligations of E1 in terms of elements requiring insulation and the nature of insulation required.

It will be seen from Table 5.1 that the criterion applied to assess airborne sound insulation is a single figure index, $D_{nT,w} + C_{tr}$, which is obtained from a frequency band analysis of the noise. The term 'airborne sound insulation' is explained in Chapter 3 and the procedure required for measurement of $D_{nT,w} + C_{tr}$ is described in Chapter 6. This index replaces $D_{nT,w}$ which applied in the 1992 approved document by the addition of the C_{tr} term. C_{tr} modifies the index by taking greater account of low frequency noise. This is considered important since, for example, the powerful low frequency output of modern musical entertainment systems is such that sound transmission at low frequencies is now a frequent source of considerable annoyance.

The value of 45dB for airborne sound insulation between purpose-built dwellings which appears in Table 5.1 appears at first sight to represent a reduction in standard set when compared with the equivalent value of 49dB which appears in the preceding code of practice. However, the effect of C_{tr} will be to reduce the measured value by approximately 5dB and, in addition, a reduction of 2dB is made to take into account the accuracy of the field measurement procedure. Consequently, the new value may be considered as an increase in the standard of 3dB, i.e:

$$49dB - 5dB - 2dB + 3dB = 45dB$$

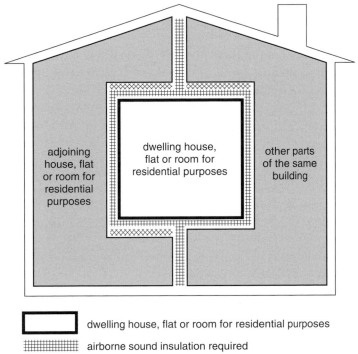

dwelling house, flat or room for residential purposes

airborne sound insulation required

impact sound insulation required

n.b. airborne sound insulation is also required where walls separate dwelling houses, flats and rooms for residential purposes from adjoining buildings and refuse chutes.

Fig. 5.1 Requirements for protection against sound between dwelling places.

In the case of a construction with a particularly poor low frequency performance, the overall improvement in performance would need to be higher than 3dB in order to compensate for the lowering effect of C_{tr}.

Required values of the index used to assess the standard of impact sound insulation, $L'_{nT,w}$, are shown in Table 5.1. See Chapter 4 for explanations of the term 'impact sound insulation' and the index $L'_{nT,w}$, and Chapter 6 for an explanation of the associated measurement procedure.

There are two sets of circumstances when it may not be necessary to fully comply with the requirement E1. Firstly, buildings constructed from prefabricated assemblies which are newly made or delivered from stock are subject to the full requirements of Schedule 1 of the Building Regulations but in some cases, e.g. temporary dwelling places, the requirement to provide reasonable sound insulation may vary. Examples referred to in the approved document are buildings:

- Formed by dismantling, transporting and reassembling sub-assemblies on the same premises

- Constructed from sub-assemblies from other premises or stock manufactured before 1 July 2003.

These buildings would normally be acceptable if they satisfied the requirements of Approved Document E 1992 or, in the case of school buildings, the relevant provisions of the 1997 edition of Building Bulletin 87 [13].

Secondly, in the case of the material change of use of some historic buildings it may not be viable to fully comply with E1 whilst still conserving their special historical features. In such cases the aim should be to improve sound insulation as much as is practically possible without adversely affecting the character of the building or its long-term rate of deterioration. When this approach is adopted, and E1 is not fully complied with, a prominent notice should be displayed in the building indicating the level of insulation obtained by tests carried out using the appropriate procedure. For the purposes of Approved Document E 'historic buildings' include:

- Listed buildings
- Buildings in conservation areas
- Buildings of architectural or historic interest and referred to as a material consideration in local authority development plans or in national parks, areas of outstanding national beauty or world heritage sites.

It is suggested in the AD that when assessing the relative merits of reduced sound transmission and conservation, advice should be obtained from the local planning authority's conservation officer.

5.2   Sound insulation within dwelling places   AD E: 0.9–0.10

Requirement E2 of Schedule 1 to the Building Regulations 2000 (as amended) states that the design and construction of houses, flats and rooms for residential purposes should be such that:

- internal walls between bedrooms and any other rooms
- internal walls between rooms containing water closets and any other rooms
- all internal floors

provide reasonable resistance to the transmission of sound. These requirements are shown diagrammatically in Fig. 5.2.

Requirement E2 will be satisfied if the above elements meet the insulation values quoted in Table 2 of the AD. This simply states that new internal walls and floors in dwelling houses, flats and rooms for residential purposes, whether purpose-built or formed by material change of use, must have a value of airborne sound insulation, R_w = 40dB or above. R_w, the weighted sound reduction index (see Chapter 3), is a single figure index which results from a frequency band analysis carried out in a laboratory of a sample of the element of construction. It follows therefore that

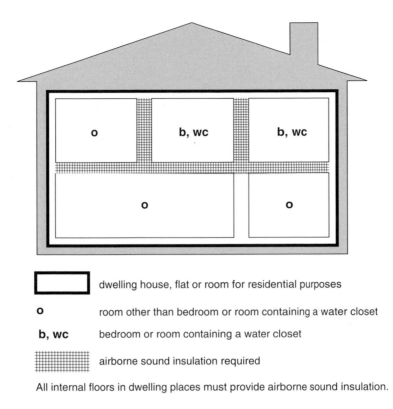

dwelling house, flat or room for residential purposes

o   room other than bedroom or room containing a water closet

b, wc   bedroom or room containing a water closet

                     airborne sound insulation required

All internal floors in dwelling places must provide airborne sound insulation.

Fig. 5.2   Requirements for protection against sound within dwelling places.

pre-completion testing of internal elements is not necessary but that builders will need to show that elements used satisfy the above criterion. For laboratory results to be acceptable it will be necessary to demonstrate that the organisation undertaking the work has appropriate accreditation, preferably by the United Kingdom Accreditation Service (UKAS), or a European equivalent, for laboratory measurements. The laboratory measurement of sound insulation is discussed later in Chapter 6. There is no requirement to provide impact sound insulation in order to satisfy E2.

As with E1, Requirement E2 extends to dwelling places resulting from the material change of use of a building but unlike E1 the same value applies to both categories of accommodation. Figure 5.2 summarises the obligations of E2 in terms of elements requiring insulation.

5.3   Reverberation in common spaces and acoustic conditions in schools   AD E: 0.11–0.12

Requirement E3 of the Building Regulations 2000 (as amended) states that the construction of the common internal spaces in buildings which contain flats or rooms for residential purposes shall be such as to prevent more reverberation than is reasonable. This requirement will be satisfied if sound absorptive measures described in section 7 of the approved document are applied. The provision of absorption in common spaces is considered in Chapter 12.

Requirement E4 states that the design and construction of all rooms and other places in school buildings shall be such that their acoustic conditions and insulation against disturbance by noise is appropriate to their intended use. It is stated in AD E that the normal way of satisfying these conditions will be to meet the requirements given in section 1 of Building Bulletin 93, *The acoustic design of schools* [2]. See Chapter 14.

6 Measurement Procedures

6.1 Types of measurement

There are several types of measurement relevant to AD E. A general distinction may be made between:

- Laboratory measurements – measurements carried out in a test facility specifically designed for the purpose, usually for the purpose of determining the potential performance of a building element or elements
- Field measurements – measurements carried out in completed buildings, usually for the purpose of determining the actual performance of a building element and all its associated structure and components.

The actual measurements carried out in laboratory and field tests are in essence the same, though there are some procedural differences. The measurements themselves include:

- Sound pressure level
- Reverberation time
- Airborne sound transmission
- Impact sound transmission
- Reduction of impact sound transmission
- Dynamic stiffness.

6.2 Sound pressure level

The measurement of the sound pressure level in a space is fundamental, and the main points to be taken into account are:

- The accuracy of the microphone and its associated equipment over the relevant frequency range
- Variations in the sound pressure level with time
- The precision of the filter sets which are used to give third octave band sound pressure levels

- The difference between the sound pressure levels of the generated noise and any background noise
- Variations of the sound pressure level with position in the space.

The first three of these are general to all sound pressure level measurements, and minimum performance specifications are to be found in various standards (e.g. BS EN 61094-2:1994, BS EN 61260:1996 and BS EN 61672-1:2003). However, the other two are particularly relevant to AD E. The sound pressure level of any unwanted background noise must be significantly less than the sound pressure level of the noise signal that it is desired to measure, otherwise the precision of the measurement will be compromised. This applies to both laboratory and field measurements, but is particularly relevant to the latter. For measurements of airborne and impact sound transmission, the requirement (see BS EN ISO 140-3, 140-4, 140-5 and 140-6) is that the difference should always be at least 6dB, and preferably more than 15dB for laboratory measurements or more than 10dB for field measurements. If the difference is between 6 and 15dB for laboratory measurements, or between 6 and 10dB for field measurements, a correction must be applied to remove the effect of the background noise from the measurement of sound pressure level of the signal:

$$L_s = 10 \log_{10} \left(10^{L_{sb}/10} - 10^{L_b/10} \right)$$

where L_s is the true sound pressure level of the signal
        L_{sb} is the sound pressure level of the signal plus background noise
        L_b is the sound pressure level of the background noise alone.

The presence of background can be important in field tests of airborne sound reduction.

Variations in the sound pressure level of a noise source with position in a space occur even when the output of the equipment that is generating the noise is kept constant. This occurs because the sound field in an enclosed space can never be perfectly diffuse, and there is always at least some residual effect due to standing waves, the more so as the frequency is reduced and the corresponding wavelength of the noise signal increased. There are two methods for obtaining an average of the sound pressure level. The simplest is to measure at several different positions, and to find the average sound pressure level on an equal energy basis:

$$L_{av} = 10 \log_{10} \left(\frac{1}{n} \sum_{i=1}^{n} 10^{L_i/10} \right)$$

where L_{av} is the average sound pressure level in the space
        L_i is the sound pressure level at a measurement point in the space
        n is the number of measurement points.

Alternatively, a single continuously moving microphone can be used, the output of which is integrated over a suitable measurement period:

$$L_{av} = 10 \log_{10} \left(\frac{1}{T} \int_{t=0}^{t=T} \left(\frac{p}{p_0} \right)^2 dt \right)$$

where p is the measured varying sound pressure in pascals
   p_0 is the reference sound pressure of 20 pascals
   T is the time over which the measurement is made.

6.3   Reverberation time

The determination of reverberation time is a part of the procedures for measuring both airborne and impact sound transmission. Reverberation time is defined as the time it takes for the sound pressure level in a space to decay by 60dB after the noise source has been switched off. It is generally neither possible nor desirable to attempt to measure a complete decay of 60dB, and it is usual to use a smaller range. Details are given in BS EN ISO 354:2003 and BS EN ISO 3382:2000, but the main points to note are:

- Although less than 60dB, the range of the decay should be large enough (normally at least 20dB) to give a reliable measurement
- The decay must follow, at least approximately, a straight line
- The measurable section of the decay must finish above the level of any background noise by a reasonable margin, usually at least 10dB
- For each frequency or frequency band, measurements must be repeated a minimum number of times and at several positions in the space, and the average taken.

6.4   The measurement of airborne sound transmission

The laboratory measurement of airborne sound transmission is carried out in order to test the probable performance of a separating wall or floor when installed in a building. Figure 6.1 is a schematic drawing of a typical laboratory set-up.

The measurement procedure follows the principles described in Chapter 3, that is, sound pressure levels are measured on either side of the test element, and the difference taken. The reverberation time is measured, and a correction applied for the size of the test element and the sound absorption of the receiving room. Depending on which correction is used, any of the three principal indices, R, D_n or D_{nT} may be obtained. The process is repeated at each of the 16 third octave bands from 100Hz to 3150Hz inclusive. Details of the measurement procedure are specified in BS EN ISO 140-3:1995.

As the test facility is designed to minimise flanking transmission, the result is as close as possible to the true sound reduction performance of the separating element

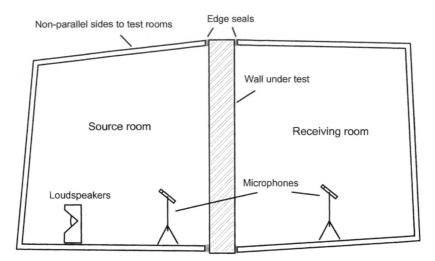

Fig. 6.1 Laboratory measurement of airborne sound transmission.

itself. For example, if the sound reduction index has been found, it will be closer to R than to the apparent sound reduction index R'. Consequently, the result of a laboratory test cannot be used by itself as proof of compliance with the requirements of AD E.

In order to relate the laboratory performance of a separating wall or floor more closely to its field performance, there is the possibility of testing in a flanking laboratory (AD E: B3.10–B3.14). In this, the test element is constructed complete with any associated structural elements that may contribute to sound transmission between the test spaces. Flanking laboratory measurements are a useful way of assessing the probable performance in practice of novel or alternative constructions and it is stated in the approved document that if a test construction provides airborne insulation, $D_{nT,w} + C_{tr}$, value of 49dB or more when measured in a flanking laboratory, this may be taken as an indication that a construction which is identical in all significant details may achieve a $D_{nT,w} + C_{tr}$ value of 45dB or more if built in the field. However, although tests in flanking laboratories are very useful because they include the effects of transmission through both direct and flanking routes, they may not by themselves be used to indicate compliance with requirement E1 since this relates to field performance.

A standard for the laboratory measurement of flanking transmission is being developed and construction details of a flanking laboratory are available from: The Acoustics Centre, BRE, Garston, Watford WD25 9XX.

The field measurement of airborne sound transmission is carried out in order to determine the actual performance of a separating wall or floor when installed in a building. The test arrangement is shown in Fig. 6.2

In this case, there is transmission of sound via flanking routes as well as through the separating wall, and the result of the measurement is the sound reduction index of the complete structure. Although essentially the same as for a laboratory

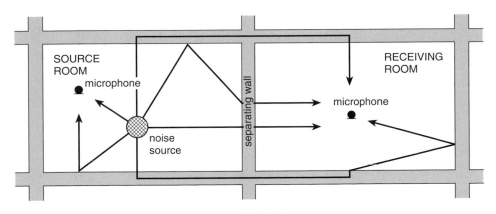

Fig. 6.2   Field measurement of airborne sound insulation.

measurement, the measurement details differ in some respects, the required procedures being specified in BS EN ISO 140-4:1998. For compliance with requirement E1, the index which must be obtained is the weighted standardised level difference, $D_{nT,w}$ (as defined in Chapter 3) and so the results must be processed to give the standardised level difference D_{nT} using:

$$D_{nT} = L_1 - L_2 + 10 \log_{10} \left(\frac{T}{T_0} \right) \quad \text{dB}$$

with T_0 set to the standard value of 0.5 s. This is measured for the 16 third octave bands from 100 Hz to 3150 Hz and then converted to $D_{nT,w}$ as shown by the example in section 6.6 below.

6.5   The measurement of impact sound transmission

A standard for the laboratory measurement of the impact sound insulation of floors is given by BS EN ISO 140-6:1998. However, the transmission of impact sound depends so strongly on the structural connections within a building that a laboratory measurement of impact sound transmission on an isolated element is unlikely to be a reliable means for predicting the impact performance of that element in an actual building. For a laboratory test to be meaningful, it is necessary to include all the relevant structure associated with the test specimen, and so it is preferable for the measurements to be carried out in a specially designed flanking laboratory, as described in section 6.4 above.

The field measurement of impact sound insulation follows the principles described in Chapter 4, and the standard to be followed is described in BS EN ISO 140-7:1998. Initially a tapping machine is located on top of the separating floor, as shown in Fig. 6.3.

The tapping machine, which must be designed and constructed in accordance with

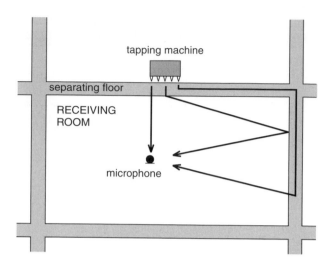

Fig. 6.3   Measurement of impact sound transmission.

the standard, is a device with five small hammers that consecutively strike the floor producing impacts of predetermined momentum. On activation of the tapping machine, the resulting sound pressure level, L_{rec}, is measured in the room beneath the floor (the receiving room). As with airborne sound transmission, the level measured at the microphone in the receiving room will be influenced by noise energy reflected within the room, and so the reverberation time is also measured when assessing impact sound insulation. For compliance with requirement E1, the index which must be obtained is the weighted standardised impact sound pressure level, $L'_{nT,w}$ (as defined in Chapter 4), and so the results must be processed to give the standardised impact sound pressure level L'_{nT}:

$$L'_{nT} = L_{rec} - 10\log_{10}\left(\frac{T}{T_0}\right)\quad\text{dB}$$

with T_0 set to the standard value of 0.5 s. This is measured for the 16 third octave bands from 100Hz to 3150Hz and then converted to $L'_{nT,w}$ as shown by the example in section 6.6 below.

The procedures for determining the reduction of impact sound due to a floor covering are essentially the same as above. In the laboratory measurement procedure, which is given in BS EN 140-8, a reinforced concrete floor slab of specified standard thickness and area (120 mm thick and at least 10 m^2 surface area) is installed between the source room and the receiving room below. The standard tapping machine is applied to the floor, and the normalised impact sound pressure level without the floor covering is obtained in third octave bands, using (see section 4.1):

$$L_{n,r,0} = L_{rec} + 10\log_{10}\left(\frac{A}{A_0}\right)\quad\text{dB}$$

where $L_{n,r,0}$ is the normalised impact sound pressure level of the reference floor
        L_{rec} is the measured sound pressure level in the receiving room
        A is the total sound absorption of the receiving room
        A_0 is the reference absorption of $10\,\text{m}^2$.

The measurements are repeated, in the same third octave bands, with the floor covering laid on the standard concrete floor to obtain:

$$L_{n,r} = I_{rec} + 10\log_{10}\left(\frac{A}{A_0}\right) \quad \text{dB}$$

where $L_{n,r}$ is the normalised impact sound pressure level of the reference floor plus floor covering.

The reduction in impact sound level, ΔL, is then found for each third octave band from:

$$\Delta L = L_{n,r,0} - L_{n,r}$$

The results for ΔL are useful in identifying the frequencies at which the floor covering is most (or least) effective. However, the ΔL values are not used to find the single number weighted reduction of impact sound level (see sections 4.3 and 6.6).

6.6   Calculation of single number quantities

The criteria used in AD E for determining compliance with the requirements of Part E are:

- The weighted standardised level difference, $D_{nT,w}$
- The weighted standardised impact sound pressure level, $L'_{nT,w}$
- The weighted reduction in impact sound pressure level, ΔL_w.

The first two of these criteria are derived from the third octave band measurements of either the standardised level difference D_{nT} or the standardised impact sound pressure level L'_{nT}, using the reference values given in Tables 3.2 and 4.1. In addition, for airborne sound only, it is necessary to calculate the spectrum adaptation term, C_{tr}, which requires data from Table 3.4. The third criterion, ΔL_w is derived from a laboratory measurement of $L_{n,r,0}$, the weighted normalised impact sound pressure level of a floor covering laid on a reference floor. The processes can be explained and illustrated by the following examples, in which it is assumed that the measurements have been completed, and results for D_{nT}, L'_{nT} and $L_{n,r}$ in the 16 third octave bands have been obtained. Examples 1, 2 and 3 illustrate the calculation of $D_{nT,w}$ and C_{tr}, and examples 4 and 5 illustrate the calculation of $L'_{nT,w}$ and ΔL_w.

Example 1 – Calculation of the weighted standardised level difference, $D_{nT,w}$, and the spectrum adaptation term, C_{tr}, for a separating wall

Field measurements to find D_{nT} for a separating wall yield the results shown in column 3 of Table 6.1. Column 4 contains the reference values (from Table 3.2), and column 5 is the difference between the reference and the measured values. The total at the bottom of column 5 is the sum of the positive differences, and as this is greater than 32dB, it is necessary to reduce the whole set of reference values in steps of 1dB at a time and repeat the calculation of the difference. In this case it is found that after a downward shift of 5dB, as shown in column 6, the sum of the differences (positive values only, column 7) is reduced to 31.7dB. This is less than 32dB. The weighted standardised level difference, $D_{nT,w}$, is now obtained by reading the value of the shifted reference curve at 500Hz. It can be seen that this is 47dB.

Table 6.1 Results and calculation of $D_{nT,w}$ for separating wall no. 1.

Band No.	Frequency Hz	D_{nT} dB	Reference values, dB	Difference dB	Shifted ref. value, dB	Revised difference, dB
1	100	25.3	33	7.7	28	2.7
2	125	23.4	36	12.6	31	7.6
3	160	28.6	39	10.4	34	5.4
4	200	31.1	42	10.9	37	5.9
5	250	36.5	45	8.5	40	3.5
6	315	38.7	48	9.3	43	4.3
7	400	43.8	51	7.2	46	2.2
8	500	47.2	52	4.8	**47**	−0.2
9	630	47.9	53	5.1	48	0.1
10	800	51.1	54	2.9	49	−2.1
11	1000	52.6	55	2.4	50	−2.6
12	1250	53.9	56	2.1	51	−2.9
13	1600	58.5	56	−2.5	51	−7.5
14	2000	61.2	56	−5.2	51	−10.2
15	2500	59.2	56	−3.2	51	−8.2
16	3150	58.4	56	−2.4	51	−7.4
			Total diff. =	83.9	New diff. =	31.7

Result $D_{nT,w}$ = 47 dB

The measured values, together with both the reference values and the reference values shifted down by 5dB, are shown in Fig. 6.4.

In order to check for compliance with requirement E1, it is also necessary to compute the spectrum adaptation term, C_{tr}. This utilises spectrum number 2 from Table 3.4. The formula for the calculation (derived from **BS EN ISO 717-1:1997**) is:

$$C_{tr} = -10 \log_{10} \left(\sum_{i=1}^{16} 10^{(L_i - D_{nT,i})/10} \right) - D_{nT,w} \quad \text{dB}$$

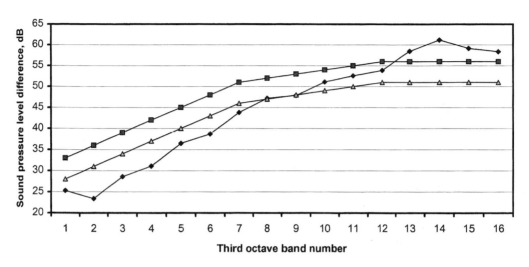

Fig. 6.4   Separating wall number 1.

where i is an integer, used to count through the 16 third octave bands
        L_i is the value of the C_{tr} spectrum at frequency band i
        $D_{nT,i}$ is the standardised level difference in frequency band i
        $D_{nT,w}$ is the weighted standardised level difference (from Table 6.1).

The calculation is shown in Table 6.2. Column 3 is the measured values of D_{nT} (the same as column 3 of Table 6.1) column 4 is the C_{tr} spectrum (i.e. L_i), and column 5 is the difference ($L_i - D_{nT,i}$). Column 6 is the antilogarithm of column 5 (multiplied by 10^5 for convenience of presentation). The total at the foot of the table is the required summation.

Substituting in the formula gives:

$$C_{tr} = -10\log_{10}\left(13.6295 \times 10^{-5}\right) - 47 = -8 \quad \text{dB}$$

The strongly negative result for C_{tr} is an indication that this separating wall has a relatively poor sound reduction performance at low frequency. Combining this with $D_{nT,w}$ gives:

$$D_{nT,w} + C_{tr} = 47 - 8 = 39 \quad \text{dB}$$

Referring to Table 5.1, the figure of 39dB is considerably less than both the 45dB and the 43dB requirements for all types of separating wall. Therefore this particular separating wall cannot comply with the requirements of E1.

Table 6.2   Calculation of C_{tr} for separating wall no. 1.

Band No.	Frequency Hz	D_{nT} dB	C_{tr} spectrum, L_i dB	Difference dB	$10^{(L-D)/10} \times 10^5$
1	100	25.3	−20	−45.3	2.9512
2	125	23.4	−20	−43.4	4.5709
3	160	28.6	−18	−46.6	2.1878
4	200	31.1	−16	−47.1	1.9498
5	250	36.5	−15	−51.5	0.7079
6	315	38.7	−14	−52.7	0.5370
7	400	43.8	−13	−56.8	0.2089
8	500	47.2	−12	−59.2	0.1202
9	630	47.9	−11	−58.9	0.1288
10	800	51.1	−9	−60.1	0.0977
11	1000	52.6	−8	−60.6	0.0871
12	1250	53.9	−9	−62.9	0.0513
13	1600	58.5	−10	−68.5	0.0141
14	2000	61.2	−11	−72.2	0.0060
15	2500	59.2	−13	−72.2	0.0060
16	3150	58.4	−15	−73.4	0.0046
				Total =	13.6295

Example 2 – Calculation of the weighted standardised level difference, $D_{nT,w}$ and the spectrum adaptation term, C_{tr}, for a separating wall with improved low frequency performance

Table 6.3 shows the results of field measurements to find D_{nT} for a separating wall with better low frequency performance and poorer high frequency performance when compared to example 1. The calculation of $D_{nT,w}$ proceeds as shown in Table 6.3. In this case, the reference curve must be shifted down by 3dB in order to bring the total of the (positive) differences below 32dB.

The calculation of C_{tr} is shown in Table 6.4.

Substituting in the formula gives:

$$C_{tr} = -10\log_{10}\left(4.0716 \times 10^{-5}\right) - 49 = -5 \quad \text{dB}$$

The negative result for C_{tr} is not as strong as for separating wall no. 1, due to the better low frequency performance. Combining this with $D_{nT,w}$ gives:

$$D_{nT,w} + C_{tr} = 49 - 5 = 44 \quad \text{dB}$$

At 44dB this wall would be acceptable for a purpose built room for residential purposes or in a property formed by material change of use. It does not comply with requirement E1 for purpose built houses or flats.

Table 6.3 Results and calculation of $D_{nT,w}$ for separating wall no. 2.

Band No.	Frequency Hz	D_{nT} dB	Reference values, dB	Difference dB	Shifted ref. value, dB	Revised difference, dB
1	100	33.3	33	−0.3	30	−3.3
2	125	30.4	36	5.6	33	2.6
3	160	34.6	39	4.4	36	1.4
4	200	36.1	42	5.9	39	2.9
5	250	40.5	45	4.5	42	1.5
6	315	41.7	48	6.3	45	3.3
7	400	45.8	51	5.2	48	2.2
8	500	48.2	52	3.8	**49**	0.8
9	630	46.9	53	6.1	50	3.1
10	800	49.1	54	4.9	51	1.9
11	1000	49.6	55	5.4	52	2.4
12	1250	49.9	56	6.1	53	3.1
13	1600	53.5	56	2.5	53	−0.5
14	2000	55.2	56	0.8	53	−2.2
15	2500	52.2	56	3.8	53	0.8
16	3150	50.4	56	5.6	53	2.6
			Total diff. =	70.9	New diff. =	28.6

Result $D_{nT,w}$ = 49 dB

Table 6.4 Calculation of C_{tr} for separating wall no. 2.

Band No.	Frequency Hz	D_{nT} dB	C_{tr} spectrum, L_i dB	Difference dB	$10^{(L-D)/10} \times 10^{-5}$
1	100	33.3	−20	−53.3	0.4677
2	125	30.4	−20	−50.4	0.9120
3	160	34.6	−18	−52.6	0.5495
4	200	36.1	−16	−52.1	0.6166
5	250	40.5	−15	−55.5	0.2818
6	315	41.7	−14	−55.7	0.2692
7	400	45.8	−13	−58.8	0.1318
8	500	48.2	−12	−60.2	0.0955
9	630	46.9	−11	−57.9	0.1622
10	800	49.1	−9	−58.1	0.1549
11	1000	49.6	−8	−57.6	0.1738
12	1250	49.9	−9	−58.9	0.1288
13	1600	53.5	−10	−63.5	0.0447
14	2000	55.2	−11	−66.2	0.0240
15	2500	52.2	−13	−65.2	0.0302
16	3150	50.4	−15	−65.4	0.0288
				Total =	4.0716

Example 3 Calculation of the weighted standardised level difference, $D_{nT,w}$, and the spectrum adaptation term, C_{tr}, for a separating wall with improved overall performance

Table 6.5 shows the results of field measurements to find D_{nT} for a separating wall with the low frequency performance improved still further. The calculation proceeds as before, with the reference values being shifted down by 2dB to reduce the adverse deviation below 32dB.

Table 6.5   Results and calculation of $D_{nT,w}$ for separating wall no. 3.

Band No.	Frequency Hz	D_{nT} dB	Reference values, dB	Difference dB	Shifted ref. value, dB	Revised difference, dB
1	100	34.7	33	−1.7	31	−3.7
2	125	32.6	36	3.4	34	1.4
3	160	36.1	39	2.9	37	0.9
4	200	37.8	42	4.2	40	2.2
5	250	41.1	45	3.9	43	1.9
6	315	42.7	48	5.3	46	3.3
7	400	46.0	51	5.0	49	3.0
8	500	48.2	52	3.8	**50**	1.8
9	630	47.5	53	5.5	51	3.5
10	800	49.7	54	4.3	52	2.3
11	1000	50.2	55	4.8	53	2.8
12	1250	50.5	56	5.5	54	3.5
13	1600	53.6	56	2.4	54	0.4
14	2000	55.3	56	0.7	54	−1.3
15	2500	52.3	56	3.7	54	1.7
16	3150	50.9	56	5.1	54	3.1
			Total diff. =	60.5	New diff. =	31.8

Results $D_{nT,w}$ = 50 dB

The calculation of C_{tr} is shown in Table 6.6.
Substituting in the formula gives:

$$C_{tr} = -10\log_{10}\left(3.0370 \times 10^{-5}\right) - 50 = -5 \quad \text{dB}$$

Despite an improved low frequency performance, the negative result for C_{tr} is the same as for separating wall no. 2. Combining this with $D_{nT,w}$ gives:

$$D_{nT,w} + C_{tr} = 50 - 5 = 45 \quad \text{dB}$$

This just satisfies the requirement of 45dB for separating walls in purpose built houses and flats.

Table 6.6 Calculation of C_{tr} for separating wall no. 3.

Band No.	Frequency Hz	D_{nT} dB	C_{tr} spectrum, L_i dB	Difference dB	$10^{(L-D)/10} \times 10^{-5}$
1	100	34.7	−20	−54.7	0.3388
2	125	32.6	−20	−52.6	0.5495
3	160	36.1	−18	−54.1	0.3890
4	200	37.8	−16	−53.8	0.4169
5	250	41.1	−15	−56.1	0.2455
6	315	42.7	−14	−56.7	0.2138
7	400	46.0	−13	−59.0	0.1259
8	500	48.2	−12	−60.2	0.0955
9	630	47.5	−11	−58.5	0.1413
10	800	49.7	−9	−58.7	0.1349
11	1000	50.2	−8	−58.2	0.1514
12	1250	50.5	−9	−59.5	0.1122
13	1600	53.6	−10	−63.6	0.0437
14	2000	55.3	−11	−66.3	0.0234
15	2500	52.3	−13	−65.3	0.0295
16	3150	50.9	−15	−65.9	0.0257
				Total =	3.0370

Example 4 – Calculation of the weighted standardised impact sound pressure level, $L'_{nT,w}$, for a separating floor

Table 6.7 shows the results of field measurements to find L'_{nT} on a separating floor, and the calculations to find $L'_{nT,w}$.

Table 6.7 Results and calculation of $L'_{nT,w}$ for a separating floor.

Band No.	Frequency Hz	L'_{nT} dB	Reference values, dB	Difference dB	Shifted ref. value, dB	Revised difference, dB
1	100	73.3	62	11.3	67	6.3
2	125	75.6	62	13.6	67	8.6
3	160	72.4	62	10.4	67	5.4
4	200	71.5	62	9.5	67	4.5
5	250	70.1	62	8.1	67	3.1
6	315	68.9	62	6.9	67	1.9
7	400	66.8	61	5.8	66	0.8
8	500	66.0	60	6.0	**65**	1.0
9	630	63.6	59	4.6	64	−0.4
10	800	62.4	58	4.4	63	−0.6
11	1000	60.1	57	3.1	62	−1.9
12	1250	56.2	54	2.2	59	−2.8
13	1600	51.8	51	0.8	56	−4.2
14	2000	47.1	48	−0.9	53	−5.9
15	2500	44.7	45	−0.3	50	−5.3
16	3150	39.8	42	−2.2	47	−7.2
			Total diff. =	86.7	New diff. =	31.6

The measured values, together with both the reference values and the reference values shifted up by 5dB, are shown in Fig. 6.5.

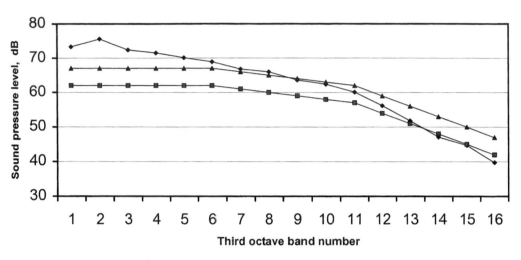

Fig. 6.5   Separating floor, example 4.

The difference (adverse deviation) is 86.7dB using the reference values, and this becomes 31.6dB when the reference curve is shifted up by 5dB. The shifted reference value at 500Hz is, from the table, 65dB, and the weighted standardised impact sound pressure level is, by definition, 5dB less than this, giving:

$$L'_{nT,w} = 65 - 5 = 60 \quad \text{dB}$$

As there is no requirement for a spectrum adaptation term for impact sound, this is the final result. It is below all the maxima for impact sound insulation in Table 5.1, and so this floor complies.

Example 5 – Calculation of the weighted reduction of impact sound pressure level, ΔL_w, for a floor covering

In this case the measurements are carried out in a laboratory, and the results processed to give normalised (not standardised) values. Thus Table 6.8 shows the results of laboratory measurements to find $L_{n,r}$ for a floor covering laid on the standard concrete floor, and the calculations to find $L_{n,r,w}$ and hence to find ΔL_w.

The difference (adverse deviation) is 51.4dB using the reference values, and this becomes 30.1dB when the reference curve is shifted up by 2dB. The shifted reference value at 500Hz is, from the table, 62dB, and the weighted normalised impact sound pressure level is, by definition, 5dB less than this, giving the equation overleaf:

Table 6.8   Results and calculation of $L_{n,r,w}$ for a separating floor with floor covering.

Band No.	Frequency Hz	$L_{n,r}$ dB	Reference values, dB	Difference dB	Shifted ref. value, dB	Revised difference, dB
1	100	68.3	62	6.3	64	4.3
2	125	71.2	62	9.2	64	7.2
3	160	68.4	62	6.4	64	4.4
4	200	68.0	62	6.0	64	4.0
5	250	67.1	62	5.1	64	3.1
6	315	66.4	62	4.4	64	2.4
7	400	64.8	61	3.8	63	1.8
8	500	63.9	60	3.9	**62**	1.9
9	630	61.6	59	2.6	61	0.6
10	800	60.4	58	2.4	60	0.4
11	1000	58.2	57	1.2	59	−0.8
12	1250	54.1	54	0.1	56	−1.9
13	1600	49.5	51	−1.5	53	−3.5
14	2000	45.4	48	−2.6	50	−4.6
15	2500	42.7	45	−2.3	47	−4.3
16	3150	38.1	42	−3.9	44	−5.9
			Total diff. =	51.4	New diff. =	30.1

$$L_{n,r,w} = 62 - 5 = 57 \quad \text{dB}$$

The weighted reduction of impact sound pressure level, ΔL_w is then found from:

$$\Delta L_w = 78 - 57 = 21 \quad \text{dB}$$

To satisfy the performance criterion in AD E for certain types of floating covering, the result should be not less than 29dB, and so this covering does not comply. Note that in order to satisfy the criterion, the result must be 29dB or above for two tests, one with the floor covering loaded, and the other with it unloaded.

6.7   Dynamic stiffness

The dynamic stiffness of a resilient material or a component is relevant to:

- The effectiveness of resilient layers in floating floor construction
- The effect of wall ties on the sound reduction of cavity walls.

In both cases the technique is to place the resilient material between two layers or masses, and to force them into vibration by applying an oscillatory force. The frequency of the applied force is varied in order to identify the resonant frequency, f_r, of the system. In general, the resonant frequency found in this way depends on the degree of damping in the material or the component, and on the applied force. The degree of damping is usually an inherent property of the material or component, but the dependence on the magnitude of the force can be reduced by repeating the measurement of f_r, each measurement using a smaller force. The results are then extrapolated to zero force. The result is usually called the apparent dynamic stiffness. For a layer of resilient material to be used under a floating floor, procedural details are given in BS EN 29052-1:1992. The result is calculated from:

$$s'_t = 4\pi^2 m'_t f^2_r$$

where s'_t is the apparent dynamic stiffness per unit area
       m'_t is the mass per unit area used in the test
       f_r is the measured resonant frequency.

Corrections for airflow resistivity must be applied in certain cases.

For a component that is to be used as a wall tie, the measurement procedure described by Hall and Hopkins [14] and by Hopkins, Wilson and Craik [15] is appropriate. In this case the result is calculated from:

$$k_{X,mm} = 2\pi^2 m_{av} f^2_r$$

where $k_{X,mm}$ is the dynamic stiffness of the wall tie when the test masses are spaced X mm apart
       m_{av} is the average mass of the two test masses used in the test apparatus
       f_r is the measured resonant frequency.

6.8   Pre-completion testing

Pre-completion testing is central to AD E. This section describes the specific requirements for the content and implementation of a pre-completion test programme.

6.8.1   Requirements   AD E: 1.1–1.10

Approved document E provides guidance on the nature and extent of the pre-completion testing that should be carried out in order to comply with Regulation 20A, or 12A which states that where Requirement E1 applies, sound insulation testing must be carried out in accordance with an approved procedure.

It is written in the Approved Document that sound insulation testing should now be considered as part of the construction process, the work being carried out at the cost of the builder. Regulations 12A and 20A state that it is the responsibility of the builder to ensure that:

- The work is carried out in an approved manner
- The results are recorded in an approved manner
- The results are given to the approved authority or inspector within 5 days of the completion of the work.

However, it is expected that the building control body will actually select which properties to test and the nature of the tests in each case. Whilst Regulations 12A and 20A state that the testing is the responsibility of the person carrying out the building work, it is recognised that the testing will be carried out by an appropriate testing body. Indeed, for field measurements of sound insulation to be acceptable it will be necessary to demonstrate that the organisation undertaking the work has appropriate accreditation for field measurements, preferably by the United Kingdom Accreditation Service (UKAS), or a European equivalent.

Testing must be carried out on:

- purpose built dwelling houses, flats and rooms for residential purposes
- dwelling houses, flats and rooms for residential purposes formed by material change of use

with the standard to be achieved being that given in Table 5.1. Guidance on appropriate construction details is given in the approved document (AD E), and is described later in this book. The approved document recommends to developers that this, or other suitable guidance, is closely followed in order to ensure that the standards set in Table 5.1 are achieved. This is very important since there is no tolerance for measurement uncertainty, which has been built into the values given in the table, and if a test does not reach the prescribed value, by even a very small margin, it will be deemed to be a failure. Where there is doubt as to the required form of construction, specialist advice should be obtained at design stage. The exception to the standards defined in Table 5.1 is when historic buildings are undergoing a material change of use and it is not practicable to achieve the prescribed standard. In this case, appropriate testing is still required with, as mentioned earlier, a notice of the standard achieved posted in a prominent position in the building.

The tests are designed to be carried out on the common wall that separates specific types of rooms and to be undertaken as soon as the dwelling place is, with the exception of decoration, complete. In the case of impact tests, with certain exceptions which are referred to in section 6.9.2.5, it is important that these are carried out before soft coverings are installed.

6.8.2   Grouping for tests   AD E: 1.11–1.18

Not all constructions have to be tested but those that are tested must allow inferences to be drawn of the performance of all constructions in the particular development. It is therefore necessary to classify the units into a number of notional groups which are representative of the whole. Houses, flats and rooms for residential purposes must form three different groups.

If there are any significant differences in the method of construction used within each of these groups, then each must be classified as a sub-group. The Approved Document provides guidance, as follows, as to how sub-groups should be determined:

- Dwelling houses (including bungalows) by type of separating wall, and flats and rooms for residential purposes by type of separating wall and separating floor
- Where there are significant differences in flanking elements, e.g. walls, floors and cavities
- Units with specific features which are acoustically unfavourable.

It should be noted that sub-grouping is not necessary if houses, flats and rooms for residential purposes are built with the same separating and flanking constructions and have similar room layouts and dimensions. However, whilst the same principles apply in the case of material change of use as for new buildings, in this case it is probable that significant differences between separating walls, separating floors and flanking constructions will occur more frequently. It follows that more sub-groups will be necessary than in the case of new build work.

6.8.3   Sets of tests   AD E: 1.19–1.28

The nature of the tests required between dwelling places depends on the type of accommodation they offer, i.e. there are different requirements for houses and flats. The tests required on a specific type of dwelling place are referred to collectively as a 'set of tests' and the specific requirements for sets of tests between different types of dwelling place are shown by a tick in Table 6.9. Note that:

- It is not appropriate to undertake impact tests between houses and bungalows
- Tests are only required in the circumstances identified in the table
- The tests referred to in Table 6.9 should each be carried out between a pair of adjacent rooms on opposite sides of the separating wall or floor between them.

Each set of tests should preferably contain individual tests between bedrooms and between living rooms. However, although it is recognised that this may not always be possible when, for example, room types are not mirrored on each side of a separating wall or floor, at least one of the rooms in one pair should be a bedroom and at least one in the other pair should be a living room. Tests should not be carried out between living spaces and corridors, stairwells or halls even though the walls between them may need to be constructed to separating wall standard. When the layout of a property is such that there is only one pair of rooms opposite each other over the complete area of a separating wall or floor, the number of sound insulation tests may be reduced accordingly.

   In the case of rooms for residential purposes, whether in new buildings or constructed by material change of use, the sound insulation of their walls and floors should be tested by application of the measurement principles applicable to dwelling houses and flats but adapted to suit the individual circumstances. It is sometimes the case that dwelling places are sold before fitting out with internal walls and internal fitments, the example of loft apartments being quoted in the approved document. In these situations the dimensions and locations of the principal room are not available but tests should still be undertaken, again applying the principles outlined for

Table 6.9   Requirements for a 'set of tests' between different types of dwelling places.

	Dwelling houses (including bungalows)	Flats with separating floors but without separating walls	Flats with separating floors and separating walls
Airborne test of walls Where possible between rooms suitable for use as living rooms.	✓	–	✓
Airborne test of walls Where possible between rooms suitable for use as bedrooms.	✓	–	✓
Airborne test of floors Where possible between rooms suitable for use as living rooms.	–	✓	✓
Airborne test of floors Where possible between rooms suitable for use as bedrooms.	–	✓	✓
Impact test of floors Where possible between rooms suitable for use as living rooms.	–	✓	✓
Impact test of floors Where possible between rooms suitable for use as bedrooms.	–	✓	✓
Total number of tests in a set	2	4	6

dwelling houses and flats. It is stated in the approved document that it is necessary in such cases to make sure that the subsequent work will not be detrimental to sound insulation by application of guidance provided on relevant construction details (see the following chapters).

6.8.4   Testing programmes AD E: 1.29–1.31

It is expected that the building control authority will determine the nature and extent of the testing programme that is required. However, the approved document lays down the overall strategy that should be adopted.

Tests should be carried out between the first pair of properties scheduled for completion/sale in each group or sub-group irrespective of the size of the development, thus providing all parties with an early indication as to the likelihood of any problems. To this end the frequency of testing should be higher in the early stages of completion than towards its end, although in the case of large developments testing should continue for a substantial part of the time that construction is taking place. Notwithstanding the fact that the rate of testing may change during the construction period, the approved document states that, assuming that there are no failures, the building control body should require at least one set of tests for every ten dwelling units completed in each group or sub-group. It should be recognised that each set of tests relates to the sound insulation between at least two properties, three in the case of flats with separating walls and floors.

6.8.5 Failure of tests AD E: 1.32–1.39

A set of tests is deemed to have failed if one or more of its individual tests do not satisfy the requirements of Table 6.9. In the event of a failure, remedial action must be taken to rectify the failure between the actual rooms which have failed and also, since the separating element between those rooms may be defective, the developer must demonstrate that other rooms which share the element in question satisfy the performance criteria by:

- Further testing between the other rooms; and/or
- Application of remedial measures to the other rooms; and/or
- Showing that the failure is limited to the rooms initially tested.

If a test between two rooms fails, careful consideration must be given to other properties which incorporate similar elements and forms of flanking construction, and developers will be expected to take the action necessary to satisfy building control authorities that the required standards are being achieved in those properties. Clearly, this will present problems if by then some of the properties are occupied. In any event, it is stated in the approved document that following the failure of a set of tests, the rate of testing should be increased from what had previously been agreed until the building control body believe that the problems have been overcome.

The approved document recommends reference to BRE Information Paper IP 14/02 [16] for guidance relating to the failure of sound insulation tests. It should be recognised that failure may be due not to the construction of the separating wall but rather to excessive flanking transmission, the two causes requiring totally different types of remedial action. The reason for the failure will also indicate which of the other rooms that have not been tested are likely to fail to meet the standard. Such rooms will require remedial treatment. Whenever remedial treatment is applied, the building control body will normally call for further sound insulation testing in order that they can be satisfied with its effectiveness.

6.8.6 Recording of pre-completion test results AD E: 1.41

Regulations 20A and 12A state that the results of sound insulation testing shall be recorded in a manner approved of by the secretary of state. In order to comply with this requirement, the report on a set of tests must contain at least the following information in the order in which it is listed. The following list is quoted directly from the approved document:

(1)   Address of Building.
(2)   Type(s) of property. Use the definitions in Regulation 2: dwelling house, flat, room for residential purposes. State if the building is a historic building (see definition in the section on Requirements of this Approved Document).
(3)   Date(s) of testing.
(4)   Organisation carrying out testing, including:
    (a)   name and address;
    (b)   third party accreditation number (e.g. UKAS or European equivalent);
    (c)   name(s) of person(s) in charge of test;
    (d)   name(s) of client(s).
(5)   A statement (preferably in a table) giving the following information:
    (a)   the rooms used for each test within the set of tests;
    (b)   the measured single number quantity ($D_{nT,w} + C_{tr}$ for airborne sound insulation and $L'_{nT,w}$ for impact sound insulation) for each test within the set of tests;
    (c)   the sound insulation values that should be achieved according to the values set out in Section 0: Performance – Table 1a or 1b, and
    (d)   an entry stating 'Pass' or 'Fail' for each test within the set of tests according to the sound insulation values set out in section 0 Performance – Table 1a or 1b.
(6)   Brief details of the test, including:
    (a)   equipment;
    (b)   a statement that the test procedures in Annex B have been followed. If the procedure could not be followed exactly then the exceptions should be described and reasons given;
    (c)   source and receiver room volumes (including a statement on which rooms were used as source rooms);
    (d)   results of tests shown in tabular and graphical form for third octave bands according to the relevant part of the BS EN ISO 140 series and BS EN ISO 717 series, including:
        (i)   single number quantities and the spectrum adaptation terms;
        (ii) D_{nT} and L'_{nT} data from which the single number quantities are calculated.

6.9   Summary of the requirements for testing and test reports

This section is a complete summary of the requirements to be met, and of the standards to be adopted, in the conduct and reporting of tests as specified in AD E. For completeness, it includes references to standards already described.

6.9.1 General requirements   AD E: B1.4, B2.1 & B3.2

It is the responsibility of the person undertaking the building work to arrange for the necessary sound insulation testing to be carried out by a testing organisation with appropriate third party accreditation. The approved document states that the organisation undertaking the testing should preferably have United Kingdom Accreditation Service (UKAS) accreditation, or a European equivalent. It is also a requirement that a valid traceable calibration certificate exists for the instrumentation used for the measurement and that it has been tested within the preceding two years.

Measurements are based on the BS EN ISO 140 series of standards and rating is based on the BS EN ISO 717 series of standards, and as stated in the approved document it is important when calculating test results that rounding does not occur until required by the standard in question.

6.9.2 Field measurement of the sound insulation of separating walls and floors   AD E: 0.1

Requirement E1 is satisfied if separating elements comply with Tables 1a and 1b of the approved document following pre-completion testing (see Chapter 5 of this book). Regulations 20A and 12A apply to building work to which E1 applies and these regulations state that pre-completion testing must be carried out in accordance with an approved procedure. The approved procedure is described in Annex B2 of Approved Document E and its content is summarised below.

6.9.2.1   Airborne sound insulation of separating walls or floors AD E: B2.2–B2.8

Procedure:
Measure in accordance with BS EN ISO140-4:1998.
(all measurements and calculations carried out in one-third octave bands)
Rate performance in accordance with BS EN ISO 717-1:1997.
(in terms of the weighted standardised level difference $D_{nT,w}$ and spectrum adaptation term C_{tr})

(1) *Measurements using a single sound source*
The average sound pressure level (SPL) is measured in one-third octave bands in source and receiving rooms for each source position using either fixed microphone positions, and averaging values on an energy basis, or a moving microphone.

The differences between average source room SPL measurements in adjacent one-third octave bands should not exceed 6dB. If this criterion is not satisfied, the instrumentation should be adjusted and measurement repeated until it is. When it is satisfied, the average SPL in the receiving room and hence the level difference should be established.

The sound source must not be moved or its output level adjusted until measurements in the source and receiving rooms are complete.

The procedure should be repeated with the sound source moved to another position in the source room. At least two source positions should be used and the level differences obtained at each position should be averaged to obtain the level difference D defined in BS EN ISO 140-4:1998.

(2) *Measurements using simultaneously operating multiple sound sources*
The average SPL is measured in one-third octave bands in source and receiving rooms with the multiple sources operating simultaneously. Measurement is undertaken using either fixed microphone positions, and averaging values on an energy basis, or a moving microphone.

The differences between average source room SPL measurements in adjacent one-third octave bands should not exceed 6dB. If this criterion is not satisfied, the instrumentation should be adjusted and measurement repeated until it is. When it is satisfied, the average SPL in the receiving room and hence the level difference, *D* defined in BS EN ISO 140-4:1998, may be established.

6.9.2.2   *Impact sound transmission through separating floors*   *AD E: B2.9*

Procedure:
Measure in accordance with BS EN ISO140-7:1998.
(all measurements and calculations carried out in one-third octave bands)
Rate performance in accordance with BS EN ISO 717-2:1997.
(in terms of the weighted standardised impact sound pressure level, $L'_{nT,w}$)

6.9.2.3   *Measurement of reverberation time*   *AD E: B2.10*

In order to calculate the weighted standardised level difference $D_{nT,w}$, and the weighted standardised impact sound pressure level, $L'_{nT,w}$, measurements have to be corrected to take into account the degree of reverberation in the receiving room. This requires measurement of its reverberation time which is defined as the time taken for the SPL to decrease by 60dB after a sound source is switched off. The measurement standards referenced above refer to ISO 354 (BS EN 20354:1993) for the method of measuring reverberation time. This standard is now replaced by BS EN ISO 354:2003. However, the guidance in BS EN ISO 140-7:1998 with respect to positioning of equipment and number of measurements should be adhered to.

6.9.2.4   *Room types and sizes*   *AD E: B2.11*

The types of rooms to be used for testing are considered in section 6.8. They should have volumes of 25m^3 or more and if not, the volumes of the rooms used should be reported.

6.9.2.5   Room conditions for tests   AD E: B2.12–B2.17

The following conditions should apply:

(1)   Rooms (or available spaces in the case of properties sold before fitting out) used for testing should be completed but unfurnished.
(2)   Floors without soft coverings should be used for impact tests. Exceptions are separating floors Type 1 and structural concrete floor bases which have integral soft coverings. If soft coverings have been laid on other types of floor, these should be removed and if that is not possible at least half of the floor should be exposed, with the tapping machine (this is required for impact transmission measurement) used on the exposed part of the floor.
(3)   When a pair of rooms used for airborne sound insulation measurement are of different volumes, the sound source should be activated in the larger room.
(4)   All doors and windows should be closed and fitments such as kitchen units and cupboards on all walls should be empty and have open doors.

6.9.2.6   Accuracy of measurement AD E: B2.18–B2.19

SPL and reverberation time measurements should be to an accuracy of 0.1 dB and 0.01 s respectively.

6.9.2.7   Measurement using moving microphones AD E: B2.20–B2.21

If a moving microphone is employed, measurements should be taken with it centred at two or more positions. Due to the transient nature of the reverberation time measurement process, fixed microphone positions as opposed to moving microphones should be used when measuring reverberation times.

6.9.3   Laboratory measurement procedures   AD E: B1.2 & B3.1–B3.2

Laboratory testing may be required to establish the performance of building elements in order to comply with Requirement E2, components such as wall ties and floating floors to which Requirement E1 applies, as well as to assess the performance of proposed novel constructions. Appropriate laboratory testing procedures are described in Annex B3 of the approved document and are outlined below.

No rounding in calculations associated with sound insulation test results should take place until required by the relevant BS EN ISO 140 and BS EN ISO 717 series standards.

6.9.3.1   Floor coverings and floating floors: AD E: B3.3–B3.6

Procedure:
Test in accordance with BS EN ISO 140-8:1998. Note, however, that this standard requires the reinforced concrete test floor to have a thickness of 120 mm (with a

tolerance of -20 mm to $+40$ mm) and a preferred thickness of 140 mm, whereas paragraph B3.3 of the approved document specifies a thickness of 140 mm.

Rate performance in accordance with BS EN ISO 717-2:1997.

The approved document states that text has been omitted from BS EN ISO 140-8:1998 and that for the purposes of the document, section 6.2.1 of BS EN ISO 140-8:1998 should be disregarded and section 5.3.3 of BS EN ISO 140-7:1998 referred to instead.

BS EN ISO 140-8:1998 refers to ISO 354 (BS EN 20354:1993) for the method of measuring reverberation time. This standard is now replaced by BS EN ISO 354:2003. However, paragraph B3.5 of the approved document states that, with respect to positioning of equipment and number of required decay measurements, the guidance given in BS EN ISO 140-8:1998 should be as adhered to, and not that given in ISO 354.

When assessing Category II specimens, defined in BS EN ISO 140-8:1998 as large specimens including rigid homogeneous surface materials (e.g. floating floors), the value of ΔL_w should be measured with and without the floor covering carrying a defined load. ΔL_w is a measure of the improvement in impact sound insulation provided by installing a floor covering or floating floor over a laboratory test floor.

6.9.3.2   *Dynamic stiffness*   **AD E: B3.7–B3.8**

Procedure for resilient layer:
Measure in accordance with BS EN 29052-1:1992 (use sinusoidal signals method with no pre-compression of specimens)
Procedure for wall ties:
Measure in accordance with BRE Information Paper IP 3/01 [14]

6.9.3.3   *Airborne sound insulation of wall and floor elements*   **AD E: B3.9**

Procedure:
Measure in accordance with BS EN ISO 140-3:1995
Rate performance in accordance with BS EN ISO 717-1:1997
(determine weighted sound reduction index R_w)
Sound reduction index and weighted sound reduction index are described in Chapter 3.

6.9.3.4   *Flanking laboratory measurement*   **AD E: B3.10–B3.14**

Although tests in flanking laboratories are very useful because they include the effects of transmission through direct and flanking routes, they may not be used to indicate compliance with requirement E1 since this relates to field performance.

Flanking laboratory measurements are a useful way of assessing the probable performance in practice of novel or alternative constructions and it is stated in the approved document that if a test construction provides airborne insulation, $D_{nT,w} + C_{tr}$, value of 49dB or more when measured in a flanking laboratory, this may be taken as an indication that a construction which is identical in all significant details may achieve a $D_{nT,w} + C_{tr}$ value of 45dB or more if built in the field. Note the first paragraph of this section.

It is also stated in the approved document that if a test construction provides impact sound transmission, $L'_{nT,w}$, of 58dB or less measured in a flanking laboratory, this may be taken as an indication that a construction which is identical in all significant details may achieve an $L'_{nT,w}$ value of 62dB or less if built in the field. Note the first paragraph of this section.

A standard for laboratory measurement of flanking transmission is being developed and construction details of a flanking laboratory are available from: The Acoustics Centre, BRE, Garston, Watford WD25 9XX.

6.9.4   Test report information

6.9.4.1   Field test report information   AD E: B4.1

The information to be provided in a test report relating to tests done in compliance with Regulations 20A or 12A is itemised in section 1 of Approved Document E, see section 6.8.6 of this chapter. It is suggested in the approved document that, although not mandatory, it may be useful to also provide the following information which is quoted directly from the document:

(1)   sketches showing the layout and dimensions of rooms tested;
(2)   description of separating walls, external walls, separating floors, and internal walls and floors including details of materials used for their construction and finishes;
(3)   mass per unit area in kg/m^2 of separating walls, external walls, separating floors, and internal walls and floors;
(4)   dimensions of any step and/or stagger between rooms tested;
(5)   dimensions and position of any windows or doors in external walls.

6.9.4.2   Laboratory test report information   AD E: B4.2

It is stated that the following information, which is quoted directly from the approved document, should be provided in test reports:

(1)   Organisation conducting test, including:
    (a)   name and address;
    (b)   third party accreditation number (e.g. UKAS or European equivalent);
    (c)   name(s) of person(s) in charge of test.
(2)   Name(s) of client(s).
(3)   Date of test.
(4)   Brief details of test, including:
    (a)   equipment;
    (b)   test procedures.

(5) Full details of the construction under test and the mounting conditions.

(6) Results of test shown in tabular and graphical form for third octave bands according to the relevant part of the **BS EN ISO 140** series and **BS EN ISO 717** series, including:

(a) single number quantity and the spectrum adaptation terms;

(b) data from which the single number quantity is calculated.

7 Separating Walls and their Flanking Constructions – New Buildings

Approved Document E, Section 2: Separating walls and associated flanking constructions for new buildings

7.1 Construction forms AD E: 2.1–2.17

Section 2 of the Approved Document E gives examples of walls which should achieve the performance standards, for houses and flats, tabulated in section 0: Performance – Table 1a of the approved document and in Chapter 5 of this book. However, as is stated in the document, the actual performance of a wall will depend on the quality of its construction. The information in this section of the AD is provided only for guidance; it is not intended to be exhaustive, and alternative constructions may achieve the required standard. Designers are also advised to seek advice from other sources, such as manufacturers, regarding alternative designs, materials or products. The details in no way override the requirement to undergo pre-completion testing. For the provision of walls in 'rooms for residential purposes', see Chapter 11.

The separating wall constructions described in the approved document are divided into four types:

(1) **Type 1 Solid masonry**
Walls consisting of one leaf of brick or concrete block, plastered on both sides.
(2) **Type 2 Cavity masonry**
Walls consisting of two leaves of brick or concrete block separated by a cavity and usually plastered on both sides.
(3) **Type 3 Masonry between independent panels**
Walls consisting of a layer of brick or block with independent panels on each side.
(4) **Type 4 Framed walls with absorbent material**
Walls consisting of independent frames of, for example, timber with panels, typically of plasterboard, fixed to the outside of each.

Within each wall type, alternatives are presented in a ranking order such that the one which should offer the best sound insulation is described first.

It is very important that the junctions between a separating wall and the elements to which it is attached are carefully designed because this will have a significant influence on its overall acoustic effectiveness. Since the level of insulation achieved will be strongly dependent on the extent to which flanking transmission exists, considerable advice is given in the AD on this issue. Guidance is provided on the junctions between separating walls and the following other elements:

- External cavity walls with masonry inner leaf
- External cavity walls with timber framed inner leaf
- Solid masonry external walls
- Internal walls – masonry
- Internal walls – timber
- Internal floors – timber
- Internal floors – concrete
- Ground floors – timber
- Ground floors – concrete
- Ceiling and roof spaces
- Separating floors Types 1, 2 and 3 (these are considered with separating floors).

However, not all wall/junction combinations are considered. It is stated in the AD that if any element has a separating function, e.g. a ground floor which is also a separating floor for a basement flat, then the separating element requirements should take priority.

The following notes should be borne in mind when considering the alternatives which are presented:

(1) The mass per unit area of walls is quoted in kilograms per square metre (kg/m^2) and procedures for calculating it, as described in Annex A of the AD, are outlined in the Appendix to this book. Densities are quoted in kilograms per cubic metre and when used for calculating the mass per unit area of brick and blockwork they should take into account moisture content by using data available in CIBSE Guide A (1999). Alternatively, manufacturers' data relating to mass per unit area may be used where this is available.

(2) The guidance assumes solid blocks which do not have voids within them. The presence of voids may affect sound transmission and relevant information should be sought from manufacturers.

(3) Drylining laminates of plasterboard with mineral wool may be used whenever plasterboard is recommended. Manufacturers' advice should be obtained for all other drylining laminates. Plasterboard lining should always be fixed in accordance with manufacturers' instructions.

(4) The cavity widths recommended in the guidance are minimum values.

(5) The requirements of Building Regulations Part C (Site preparation and resistance to moisture) and Part L (Conservation of fuel and power) must be considered when applying Part E to junctions between separating walls and ground floors.

(6) The walls described in Tables 7.1, 7.3, 7.4 and 7.5 are only example constructions.

7.1.1   Wall ties in separating and external masonry walls   AD E: 2.18–2.24

The sound insulating properties of cavity walls are affected by the coupling effect of wall ties and so it is important that they are used correctly. In AD E, wall ties are defined, in accordance with BS1243:1978 'Metal ties for cavity wall construction', as Type A (butterfly ties) and Type B (double-triangle ties) and their layout should be as described in BS 5628-3:2001 *Code of practice for use of masonry*. However, BS 5628-3 limits the above tie types and spacing to prescribed combinations of cavity widths and minimum masonry leaf thicknesses. It is stated in the approved document that tie Type B should be used only in external cavity walls where tie Type A does not satisfy the requirements of Building Regulation Part A – Structure, and that tie Type B may decrease airborne sound insulation, due to flanking transmission via the external leaf, compared with tie Type A.

An option, which is of value in improving sound insulation, is to select wall ties with an appropriate dynamic stiffness for the cavity width as explained in AD E. The procedure for measuring the dynamic stiffness of wall ties is explained in BRE IP 3/01 *Dynamic stiffness of wall ties used in masonry cavity walls*. It is essential to ensure that, if wall ties are selected on the basis of satisfying dynamic stiffness criteria, they also comply with the requirements of Building Regulation A – Structure.

7.2   Design of wall Type 1 – solid masonry   AD E: 2.29–2.35

The insulation offered against airborne sound transmission by solid walls is primarily determined by their mass per unit area. Three forms of construction based on solid walls, i.e. wall Type 1, are described in the AD, and also in Table 7.1, and the document states that these should, if built correctly, comply with Requirement E1, see Table 5.1. Examples of each of the three types of wall are provided in the approved document and these are also described in Table 7.1 and shown in Fig. 7.1.

In order for the performance of solid separating walls to be satisfactory, the junctions between them and their surrounding construction elements must be designed correctly. The requirements detailed in the AD are explained in the following sections. See also section 7.7 for information relating to correct and incorrect construction procedures.

7.2.1   Junctions with external cavity walls which have masonry inner leaves
AD E: 2.36–2.39

- There are no restrictions on the outer leaf of the wall.
- Unless the cavity is fully filled with mineral wool, expanded polystyrene beads or some other suitable insulating material (the AD states that manufacturer's advice should be sought regarding alternative suitable materials), the cavity should be stopped with a flexible closer as shown in Fig. 7.2.

Table 7.1   Types and examples of solid masonry separating wall constructions.

	Wall Type		
	Type 1.1	**Type 1.2**	**Type 1.3**
Form			
Description	Dense aggregate concrete block, plastered on both faces	Dense aggregate cast in situ concrete, plastered on both faces	Brickwork, plastered on both faces
Minimum mass per unit area	415 kg/m^2 including plaster	415 kg/m^2 including plaster	375 kg/m^2 including plaster
Surface finish	13 mm plaster on both room faces	Plaster on both room faces	13 mm plaster on both room faces
Other factors	Blocks laid flat to full thickness of the wall		Bricks laid 'frog up', coursed with headers
Example construction which provides the required mass per unit area			
Thickness	215 mm block	190 mm of concrete	215 mm brick
Detail of structure	Dense aggregate concrete block, density 1840 kg/m^3	Density of concrete 2200 kg/m^3	Density of brick 1610 kg/m^3
Coursing	110 mm	n.a.	75 mm
Surface finish	13 mm lightweight plaster with a minimum mass per unit area of 10kg/m^2, on both room faces	13 mm lightweight plaster with a minimum mass per unit area of 10kg/m^2, on both room faces	13 mm lightweight plaster with a minimum mass per unit area of 10kg/m^2, on both room faces
Further information	215 mm blocks laid flat		

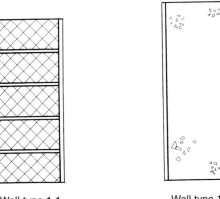

Wall type 1.1   Wall type 1.2   Wall type 1.3

Fig. 7.1   Wall Types 1.1, 1.2 and 1.3.

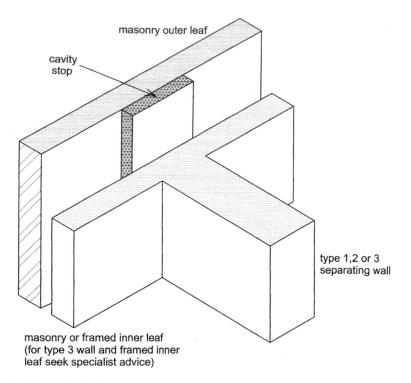

Fig. 7.2   Provision of a flexible cavity stop at junction of a Type 1, 2, or 3 separating wall and an external cavity wall.

- The separating wall should be joined to the inner leaf of the external wall either by bonding with at least half of the bond provided by the separating wall, or by using tied construction with the inner leaf abutting the separating wall, see Fig. 7.3. See also Building Regulation Part A – Structure.
- The mass per unit area of the inner leaf of the external wall should be at least 120 kg/m^2 excluding its finish in order to reduce the effects of flanking transmission. This minimum mass requirement is waived if there is no separating floor and there are openings in the external wall which are at least 1 m high, within 0.7 m of the separating wall and on both sides of the separating wall at every storey. Note however, that if openings in the external wall are placed close to and on both sides of a separating wall, there will be an increase in the sound transmitted out of the windows on one side of the seperating wall and back in through the windows on the other side. This may affect the result of a pre-completion test.

7.2.2   Junctions with external cavity walls which have timber framed inner leaves   AD E: 2.40–2.41

- There are no restrictions on the outer leaf of the wall.
- The cavity should be stopped with a flexible closer as shown in Fig. 7.2.

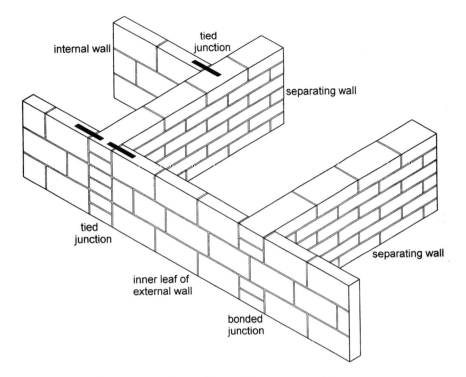

Fig. 7.3 Junctions between separating wall and other masonry wall.

- The inner leaf frame of the external wall should abut the separating wall and be tied to it vertically with ties at not greater than 300 mm centres.
- The finish of the inner leaf of the external wall should be one layer of plasterboard with a mass per unit area of at least 10 kg/m^2 and having all joints sealed with tape or caulked with appropriate sealant. Two layers of plasterboard, each with a mass per unit area of at least 10 kg/m^2, are required if there is a separating floor.

7.2.3   Junctions with solid external masonry walls   AD E: 2.42

No guidance is provided in the AD. Designers are advised to seek specialist advice.

7.2.4   Junctions with internal framed walls   AD E: 2.43

No restrictions are imposed.

7.2.5   Junctions with internal masonry walls   AD E: 2.44

Internal masonry walls should abut the separating wall and have a mass per unit area of at least 120 kg/m^2 excluding finish, see Fig. 7.3.

7.2.6 Junctions with internal timber floors   AD E: 2.45

Floor joists should not be built into this type of wall. Instead, they should be supported on joist hangers.

7.2.7 Junctions with internal concrete floors   AD E: 2.46–2.48

- The requirements for solid concrete slabs are shown in Fig. 7.4.
- Hollow core concrete plank floors and concrete beam with infill block floors should not be continuous through Type 1 separating walls.
- If concrete beam with infill block floors are built into separating walls, then the blocks in the floor must fill the space between the beams where they penetrate the wall.

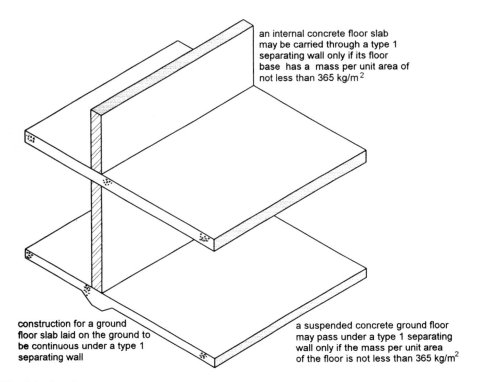

an internal concrete floor slab may be carried through a type 1 separating wall only if its floor base has a mass per unit area of not less than 365 kg/m^2

construction for a ground floor slab laid on the ground to be continuous under a type 1 separating wall

a suspended concrete ground floor may pass under a type 1 separating wall only if the mass per unit area of the floor is not less than 365 kg/m^2

Fig. 7.4   Conditions for concrete floors to pass through Type 1 separating walls.

7.2.8 Junctions with timber ground floors   AD E: 2.49

Ground floor joists should not be built into this type of wall. Instead, they should be supported on joist hangers.

7.2.9 Junctions with concrete ground floors   AD E: 2.51–2.53

- For suspended concrete ground floors and solid slabs laid directly onto the ground, see Fig. 7.4.
- Hollow core concrete plank floors and concrete beam with infill block floors should not be continuous under Type 1 separating walls.

7.2.10 Junctions with ceilings and roofs   AD E: 2.55–2.59

- The separating wall should continue to the underside of the roof with a flexible closer, which is suitable for use as a fire stop, at its junction with the roof, see Fig. 7.5.
- If the roof or loft space is not a habitable room, then the mass per unit area of the separating wall may be reduced to 150 kg/m^2 above the ceiling which separates the roof or loft space from the uppermost habitable rooms. For this reduction to apply, the mass per unit area of the ceiling must be at least 10 kg/m^2 with sealed joints, see Fig. 7.5.
- If lightweight aggregate blocks with a density of less than 1200 kg/m^3 are used above ceiling level, one side of the wall should be sealed with cement paint or plaster skim.
- Cavities in external walls should be closed at their eaves since this will have the effect of reducing any sound transmission through the cavity from the roof space. A flexible material should be used for the closure with no rigid joint

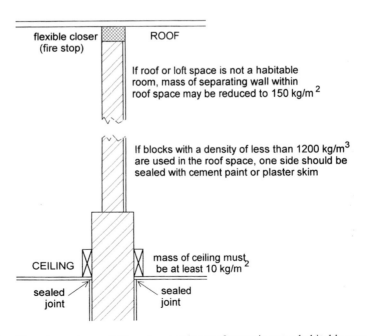

Fig. 7.5   Roof junction using wall Type 1, assuming roof space is not a habitable room.

between the two leaves. It is important to avoid a rigid connection between the two leaves and if a rigid material is used it should only be rigidly attached to one leaf, see BRE BR 262, *Thermal insulation: avoiding risks*, section 2.3 [17].

7.2.11   Junctions with separating floors   AD E: 2.60

See section 3 of the AD (Chapter 8 of this book) for details of junctions between separating floors and Type 1 separating walls.

7.3   Design of wall Type 2 – cavity masonry   AD E: 2.61–2.72

The insulation offered against airborne sound transmission by cavity masonry walls depends on the mass per unit area of the masonry, the cavity width and also the nature of the connection there is between the two leaves by, for example, wall ties and foundations.

Four types of cavity masonry separating wall are described in Approved Document E, which according to the document should, if built correctly, comply with Requirement E1, see Table 7.2. However, only two of these are appropriate for general application, two being restricted to situations where there is a step or stagger in the separating wall. Examples of each of the four types of wall are provided in the AD and these are described in Table 7.3. For each of these constructions it is important to note that:

- Cavity widths are minimum values
- Type A wall ties should be used to connect the leaves of the walls
- Blocks without voids are used (manufacturers' advice should be sought regarding the use of blocks with voids).

In order for the performance of cavity masonry separating walls to be satisfactory, the junctions between them and their surrounding construction elements must be designed correctly. The requirements detailed in the approved document are explained in the following sections. See also section 7.7 for information relating to correct and incorrect construction procedures.

7.3.1   Junctions with external cavity walls which have masonry inner leaves
AD E: 2.73–2.76

- There are no restrictions on the outer leaf of the wall.
- Unless the cavity is fully filled with mineral wool, expanded polystyrene beads or some other suitable insulating material (the approved document states that manufacturers' advice should be sought regarding alternative suitable materials), the cavity should be stopped with a flexible closer as shown in Figs. 7.2 and 7.6. Note the nature of the cavity closer for staggered wall construction in Fig. 7.6.

Table 7.2 Types of cavity masonry separating wall constructions.

	Wall Type			
	Type 2.1	Type 2.2	Type 2.3	Type 2.4
Description of wall	Two leaves of dense aggregate concrete block with 50 mm cavity, plaster on both room faces	Two leaves of lightweight aggregate block with 75 mm cavity, plaster on both room faces	Two leaves of lightweight aggregate block with 75 mm cavity and step/stagger. Plasterboard on both room faces	Two leaves of aircrete block with 75 mm cavity and step/stagger. Plasterboard or plaster on both room faces
Limitations			**Should only be used where there is a step and/or stagger of at least 300 mm**	**Should only be used in constructions without separating floors and where there is a step and/or stagger of at least 300 mm**
Minimum mass per unit area	415 kg/m² including plaster	300 kg/m² including plaster	290 kg/m² including plasterboard	150 kg/m² including finish
Minimum cavity width	50 mm	75 mm	75 mm	75 mm
Surface finish	13 mm plaster on both room faces	13 mm plaster on both room faces	Plasterboard on both room faces, minimum mass per unit area of each sheet 10 kg/m²	Plasterboard on both room faces, minimum mass per unit area of each sheet 10 kg/m² or, 13 mm plaster on both room faces
Other factors			The lightweight blocks should have a density in the range 1350 to 1600 kg/m³ The composition of the lightweight aggregate blocks contributes to performance of this construction with a plasterboard finish. Denser blocks may not give equivalent performance Increasing size of step or stagger in the separating wall tends to increase airborne sound insulation	Increasing size of step or stagger in the separating wall tends to increase airborne insulation

Table 7.3 Examples of cavity masonry separating wall constructions which provide the required mass per unit area.

	Wall Type			
	Type 2.1	**Type 2.2**	**Type 2.3**	**Type 2.4**
Description of wall	Two leaves of dense aggregate concrete block with 50 mm cavity, plaster on both room faces	Two leaves of lightweight aggregate block with 75 mm cavity, plaster on both room faces	Two leaves of lightweight aggregate block with 75 mm cavity and step/stagger. Plasterboard on both room faces	Two leaves of aircrete block with 75 mm cavity and step/stagger. Plasterboard or plaster on both room faces
Limitations			**Should only be used where there is a step and/or stagger of at least 300 mm**	**Should only be used in constructions without separating floors and where there is a step and/or stagger of at least 300 mm**
Thickness of block leaves	100 mm	100 mm	100 mm	100 mm
Description of blocks	Dense aggregate concrete block, density 1990 kg/m^3	Lightweight aggregate block, density 1375 kg/m^3	Lightweight aggregate block, density 1375 kg/m^3	Aircrete block, density 650 kg/m^3
Coursing	225 mm	225 mm	225 mm	225 mm
Surface finish	13 mm lightweight plaster with a minimum mass per unit area of 10kg/m^2, on both room faces	13 mm lightweight plaster with a minimum mass per unit area of 10kg/m^2, on both room faces	Plasterboard on both room faces. Minimum mass per unit area of each sheet 10 kg/m^2	Plasterboard on both room faces. Minimum mass per unit area of each sheet 10 kg/m^2

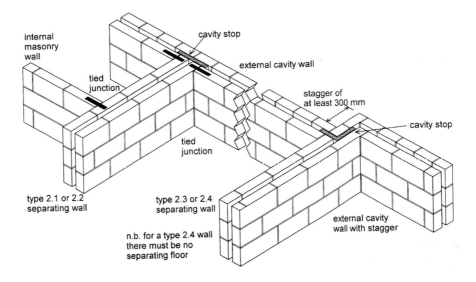

Fig. 7.6   Junctions of cavity separating walls.

- The separating wall should be joined to the inner leaf of the external wall either by bonding with at least half of the bond provided by the separating wall, or by using tied construction with the inner leaf of the external wall abutting the separating wall. See Building Regulation Part A – Structure.
- The mass per unit area of the inner leaf of the external wall should be at least 120 kg/m² excluding its finish for Type 2.2 separating walls in order to reduce the effects of flanking transmission. This requirement does not apply to Type 2.1, 2.3 and 2.4 walls unless there is also a separating floor in which case it applies to all Type 2 separating walls.

7.3.2   Junctions with external cavity walls which have timber framed inner leaves   AD E: 2.77–2.78

- There are no restrictions on the outer leaf of the wall.
- The cavity should be stopped with a flexible closer as shown in Fig. 7.2.
- The inner leaf frame of the external wall should abut the separating wall and be tied to it vertically with ties at not greater than 300 mm centres.
- The finish of the inner leaf of the external wall should be one layer of plasterboard with a mass per unit area of at least 10 kg/m² with all joints sealed with tape or caulked with appropriate sealant. Two layers of plasterboard, each with a mass per unit area of at least 10 kg/m², are required if there is a separating floor.

7.3.3 Junctions with solid external masonry walls AD E: 2.79

No guidance is provided in the AD. Designers are advised to seek specialist advice.

7.3.4 Junctions with internal framed walls AD E: 2.80

No restrictions are imposed.

7.3.5 Junctions with internal masonry walls AD E: 2.81–2.83

Internal masonry walls which abut a Type 2 separating wall, see Fig. 7.6, should have a mass per unit area of at least $120 \, \text{kg/m}^2$ excluding finish. Also, where there is a separating floor, internal masonry walls should have a mass of at least $120 \, \text{kg/m}^2$ excluding finish. In the case of Type 2.3 and 2.4 separating walls (which are only applicable if there is a step or stagger), and when there is no separating floor, there is no minimum mass requirement for internal masonry walls.

7.3.6 Junctions with internal timber floors AD E: 2.84

Floor joists should not be built into this type of separating wall. They should be supported on joist hangers.

7.3.7 Junctions with internal concrete floors AD E: 2.85

Internal concrete floors should generally be built into Type 2 separating walls and continue to the face of the cavity, see Fig. 7.7. Such floors should not bridge the cavity.

7.3.8 Junctions with timber ground floors AD E: 2.86

Ground floor joists should not be built into this type of separating wall. They should be supported on joist hangers.

7.3.9 Junctions with concrete ground floors AD E: 2.88–2.89

- Concrete floor slabs laid directly onto the ground should not be continuous beneath Type 2 separating walls but should abut them as shown in Fig. 7.7.
- Suspended concrete ground floors should not be continuous beneath Type 2 separating walls but should be built in and continue to the face of the cavity, see Fig. 7.7. Floors should not bridge the cavity.

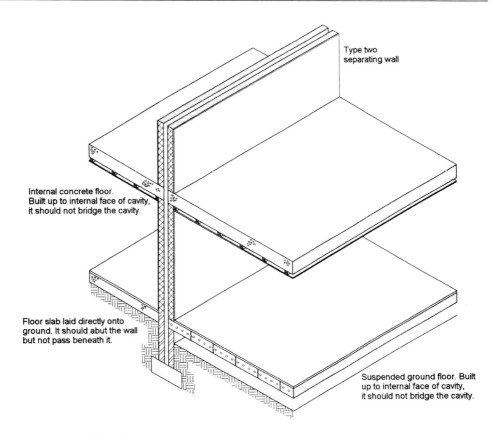

Type two
separating wall

Internal concrete floor.
Built up to internal face of cavity,
it should not bridge the cavity.

Floor slab laid directly onto
ground. It should abut the wall
but not pass beneath it.

Suspended ground floor. Built
up to internal face of cavity,
it should not bridge the cavity.

Fig. 7.7   Conditions for concrete floors and Type 2 separating walls.

7.3.10   Junctions with ceilings and roofs   AD E: 2.91–2.95

- The separating wall should continue to the underside of the roof with a flexible closer, which is suitable for use as a fire stop, at its junction with the roof, see Fig. 7.8.
- If the roof or loft space is not a habitable room, then the mass per unit area of the separating wall may be reduced to $150 \, \text{kg/m}^2$ above the ceiling which separates the roof or loft space from the uppermost habitable rooms. For this reduction to apply, the mass per unit area of the ceiling must be at least $10 \, \text{kg/m}^2$ with sealed joints and the wall must remain a cavity wall, see Fig. 7.8.
- If lightweight aggregate blocks with a density of less than $1200 \, \text{kg/m}^3$ are used above ceiling level, one side of the wall should be sealed with cement, paint or plaster skim.
- Cavities in external walls should be closed at their eaves since this will have the effect of reducing any sound transmission through the cavity from the roof space. A flexible material should be used for the closure with no rigid joint between the two leaves. It is important to avoid a rigid connection between the

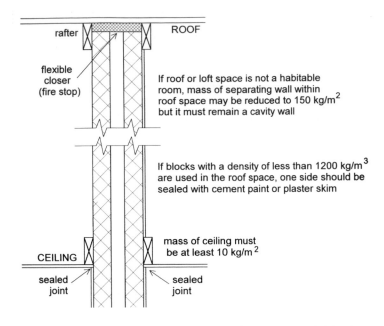

Fig. 7.8   Roof junction using wall Type 2, assuming roof space is not a habitable room.

two leaves and if a rigid material is used it should only be rigidly attached to one leaf, see BRE BR 262, *Thermal insulation: avoiding risks*, section 2.3 [17].

7.3.11   Junctions with separating floors   AD E: 2.96

See section 3 of the AD (Chapter 8 of this book) for details of junctions between separating floors and Type 2 separating walls.

7.4   Design of wall Type 3 – masonry walls between independent panels   AD E: 2.97–2.107

The independent panels used in this type of construction are typically of plasterboard fixed to timber studs and the insulation such walls offer to airborne sound transmission depends on the mass per unit area of the masonry and the independent panels. It is also influenced by the degree of isolation between the masonry core and the panels. Separating walls of this type are capable of providing very good airborne and impact sound insulation.

Three types of masonry/independent panel walls are described in Approved Document E, which according to the document should, if built correctly, comply with Requirement E1, see Table 7.4. Examples of each of the three types of wall are provided in the approved document and these are also described in Table 7.4. The essential components of this form of construction are a masonry core of either solid or

Table 7.4   Types and examples of masonry with independent panels separating wall constructions.

	Wall Type		
	Type 3.1	**Type 3.2**	**Type 3.3**
Form			
Description	Solid masonry core of dense aggregate concrete block, independent panels on both room faces	Solid masonry core of lightweight concrete block, independent panels on both room faces	Cavity masonry core of brickwork or blockwork, 50 mm cavity, independent panels on both room faces
Minimum mass per unit area of core	300 kg/m^2	150 kg/m^2	No minimum specified
Cavity width	n.a.	n.a.	50 mm minimum
Panels	Independent panels on both room faces	Independent panels on both room faces	Independent panels on both room faces
Other factors	Structural requirements determine minimum core width. Ref: Part A Structure	Structural requirements determine minimum core width. Ref: Part A Structure	Structural requirements determine minimum core width. Ref: Part A Structure
Example construction			
Thickness of core	140 mm	140 mm	Two leaves, each one at least 100 mm thick
Cavity width	n.a.	n.a.	50 mm minimum
Detail of core	Dense aggregate concrete block, density 2200 kg/m^3	Lightweight concrete block, density 1400 kg/m^3	Concrete block, density not specified
Coursing	110 mm	225 mm	Not specified
Form of panels	Independent panels, each consisting of two sheets of plasterboard with staggered joints. Mass per unit area of each panel 20 kg/m^2	Independent panels, each consisting of two sheets of plasterboard joined by a cellular core. Mass per unit area of each panel 20 kg/m^2	Independent panels, each consisting of two sheets of plasterboard joined by a cellular core. Mass per unit area of each panel 20 kg/m^2

cavity construction with independent panels on each side of it. The panels, and any framing which is associated with them, must not be in contact with the masonry core.

For each of these constructions:

- Cavity widths are minimum values;
- Type A wall ties should be used to connect the leaves of cavity masonry cores; and
- A stringent specification for the construction of the independent panels should be adhered to as follows:
    (a) the independent panels must each have a mass per unit area of at least 20 kg/m^2 excluding the mass of any supporting framework;

(b) panels to consist of two or more layers of plasterboard with staggered joints, or composite panels formed by two sheets of plasterboard separated by a cellular core; and,

(c) a gap of at least 35 mm between panels and masonry core if the panels are not supported on a frame, or a gap of at least 10 mm between the frame and the masonry core if the panels are supported on a frame.

In order for the performance of masonry with independent panel separating walls to be satisfactory, the junctions between them and their surrounding construction elements must be designed correctly. The requirements detailed in the AD are explained in the following sections. See also section 7.7 for information relating to correct and incorrect construction procedures.

7.4.1 Junctions with external cavity walls which have masonry inner leaves AD E: 2.108–2.112

- There are no restrictions on the outer leaf of the wall.
- Unless the cavity is fully filled with mineral wool, expanded polystyrene beads or some other suitable insulating material (the approved document states that manufacturers' advice should be sought regarding alternative suitable materials), the cavity should be stopped with a flexible closer as shown in Figs. 7.2 and 7.9.
- There should be a bonded or tied connection between the core of the separating wall and the inner leaf of the external wall. See Fig. 7.9 for detail of tied connection.

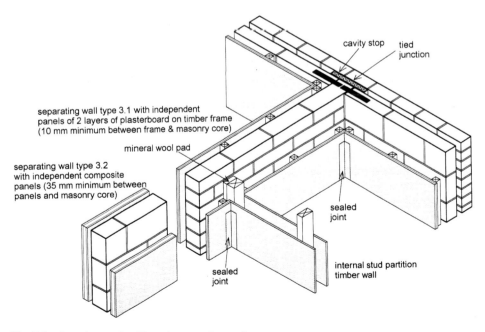

Fig. 7.9 Junctions using Type 3 separating walls.

- The inner leaf of the external wall should be lined in the same way as the separating wall, see Fig. 7.9.
- The mass per unit area of the inner leaf of the external wall should be at least $120\,kg/m^2$ excluding its finish, but if there is no separating floor and the inner leaf of the external wall is lined with independent panels in the same way as the separating wall there is no minimum mass requirement for the inner leaf.
- Where the mass per unit area of the inner leaf of the external wall is at least $120\,kg/m^2$ excluding its finish, there is no separating floor and wall Type 3.1 or 3.3 is used, the inner leaf of the external wall may be finished with plaster or plasterboard with a mass per unit area of not less than $10\,kg/m^2$.

7.4.2   Junctions with external cavity walls which have timber framed inner leaves   AD E: 2.113

No guidance is provided in the AD. Designers are advised to seek specialist advice.

7.4.3   Junctions with solid external masonry walls   AD E: 2.114

No guidance is provided in the AD. Designers are advised to seek specialist advice.

7.4.4   Junctions with internal framed walls   AD E: 2.115–2.117

Load bearing internal framed walls should be fixed to the masonry core using a continuous pad of mineral wool as shown in Fig. 7.9, whereas non-load bearing internal walls should be butted to the independent panels. The joints between all internal wall panels should be either sealed with tape or caulked with appropriate sealant.

7.4.5   Junctions with internal masonry walls   AD E: 2.118

Internal walls of masonry construction should not abut Type 3 separating walls.

7.4.6   Junctions with internal timber floors   AD E: 2.119–2.120

Floor joists should not be built into this type of wall. They should be supported on joist hangers and the spaces at the wall surface between the joists should be sealed to the full depth of the joists with timber blocking.

7.4.7   Junctions with internal concrete floors   AD E: 2.121–2.122

For separating wall Types 3.1 and 3.2 (solid), internal concrete floor slabs may only be carried through the solid masonry core if the floor base has a mass per unit area of $365\,kg/m^2$ or more. For separating wall Type 3.3 (cavity), internal concrete floors should generally be built into the wall and continue to the face of the cavity. Such floors should not bridge the cavity.

7.4.8 Junctions with timber ground floors   AD E: 2.123–2.124

Floor joists should not be built into this type of wall. They should be supported on joist hangers and the spaces at the wall surface between the joists should be sealed to the full depth of the joists with timber blocking.

7.4.9 Junctions with concrete ground floors   AD E: 2.127–2.131

- A concrete floor slab laid on the ground may be continuous under the solid core of a Type 3.1 or 3.2 (solid core) separating wall.
- A suspended concrete floor may pass under the masonry core of a Type 3.1 or 3.2 (solid core) separating wall only if the mass per unit area of the floor is 365 kg/m^2 or more.
- Hollow core concrete plank floors and concrete beam with infill block floors should not be continuous under the masonry core of a Type 3.1 or 3.2 (solid core) separating wall.
- Concrete floor slabs laid directly onto the ground should not be continuous beneath Type 3.3 (cavity core) separating walls but should abut them in a manner similar to that shown in Fig. 7.7 for cavity walls.
- Suspended concrete ground floors should not be continuous beneath Type 3.3 (cavity core) separating walls but should be built in and continue to the face of the cavity in a manner similar to that shown in Fig. 7.7 for cavity walls. The floor should not bridge the cavity.

7.4.10 Junctions with ceilings and roofs   AD E: 2.133–2.139

- The masonry core should continue to the underside of the roof with a flexible closer, which is also suitable for use as a fire stop, at its junction with the roof. Also, the independent panels should be either sealed with tape or caulked with appropriate sealant at their junction with the ceiling, see Fig. 7.10.
- Cavities in external walls should be closed at their eaves since this will have the effect of reducing any sound transmission through the cavity from the roof space. A flexible material should be used for the closure with no rigid joint between the two leaves. It is important to avoid a rigid connection between the two leaves and if a rigid material is used it should only be rigidly attached to one leaf, see BRE BR 262, *Thermal insulation: avoiding risks*, section 2.3 [17].
- If the roof or loft space is not a habitable room and the mass per unit area of the ceiling is at least 10 kg/m^2 and it has sealed joints, then the independent panels may be omitted in the roof space. Also, in the case of wall Types 3.1 and 3.2 the mass per unit area of the separating wall may be reduced to 150 kg/m^2 above the ceiling, see Fig. 7.10. However, in the case of wall Type 3.3, the cavity masonry core must continue to the underside of the roof.
- If lightweight aggregate blocks with a density of less than 1200 kg/m^3 are used above ceiling level, one side of the wall should be sealed with cement paint or plaster skim.

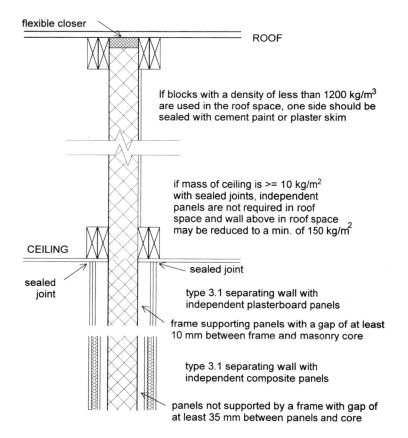

flexible closer

ROOF

If blocks with a density of less than 1200 kg/m^3 are used in the roof space, one side should be sealed with cement paint or plaster skim

if mass of ceiling is >= 10 kg/m^2 with sealed joints, independent panels are not required in roof space and wall above in roof space may be reduced to a min. of 150 kg/m^2

CEILING

sealed joint

sealed joint

type 3.1 separating wall with independent plasterboard panels

frame supporting panels with a gap of at least 10 mm between frame and masonry core

type 3.1 separating wall with independent composite panels

panels not supported by a frame with gap of at least 35 mm between panels and core

Fig. 7.10   Roof junction using wall Type 3.1 or 3.2, assuming roof space is not a habitable room.

7.4.11   Junctions with separating floors   AD E: 2.140

See section 3 of the AD (Chapter 8 of this book) for details of junctions between separating floors and Type 3 separating walls.

7.5   Design of wall Type 4 – framed walls with absorbent material AD E: 2.141–2.147

Walls of this type consist of two independent frames of timber or steel with panels fixed to the outside of each, and sound absorbent material such as mineral wool placed in the void between the two panels. The insulation they offer to airborne sound transmission depends on the mass per unit area of the panels, the extent to which the frames are isolated from each other and on the absorption properties of the void between them.

One type of framed wall consisting of timber frames and plasterboard lining is described in Approved Document E, see Table 7.5, and its construction form is

Table 7.5   One type of framed wall with absorbent material.

Form	Wall Type 4.1
Structure	A two leaf frame
Lining	Each lining to consist of two or more layers of plasterboard with staggered joints. Minimum mass per unit area of each sheet: 10kg/m^2
Width	Distance between inside lining faces: 200 mm minimum
Form of absorbent material	Unfaced mineral wool batts or quilt with a minimum density of 10kg/m^3. These may be wire reinforced
Thickness of absorbent material	25 mm min. if suspended in cavity between frames 50 mm min. if fixed to one frame 25 mm min. per batt or quilt if one is fixed to each frame
Other information	Plywood sheathing may be used in the cavity as necessary for structural reasons

Notes: 1 See Fig. 7.11 for details associated with this type of wall.
2 A masonry core may be incorporated, if this is necessary for structural reasons, but it may only be connected to one frame.

shown in Fig. 7.11. According to the document this wall should, if built correctly, comply with Requirement E1. Designers are advised in the AD to seek advice from manufacturers for steel framed alternatives.

In order for the performance of frame and absorbent infill separating walls to be satisfactory, the junctions between them and their surrounding construction elements must be designed correctly. The requirements detailed in the AD are explained in the following sections. See also section 7.7 for information relating to correct and incorrect construction procedures.

7.5.1   Junctions with external cavity walls which have masonry inner leaves   AD E: 2.148

No guidance is provided in the approved document. Designers are advised to seek specialist advice.

7.5.2   Junctions with external cavity walls which have timber framed inner leaves   AD E: 2.149–2.150

- No restrictions are imposed on the form of the outer leaf.
- The cavities in the external wall and the separating wall should be stopped with flexible closers as shown in Fig. 7.11.
- The finish of the inner leaf of the external wall, assuming there is no separating floor, should consist of one layer of plasterboard with a mass per unit area of at least $10\,\text{kg/m}^2$, all joints being sealed with tape or caulked with appropriate sealant. If there is a separating floor, two layers of plasterboard, each with a mass per unit area of at least $10\,\text{kg/m}^2$, should be provided.

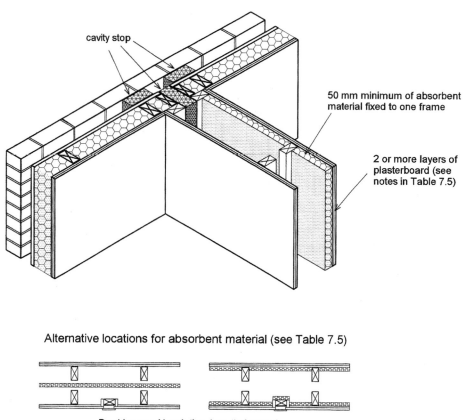

cavity stop

50 mm minimum of absorbent material fixed to one frame

2 or more layers of plasterboard (see notes in Table 7.5)

Alternative locations for absorbent material (see Table 7.5)

Provide sound insulation (e.g. timber and plasterboard) behind and around socket box.

Fig. 7.11 Details of timber framed separating wall with absorbent infill.

7.5.3   Junctions with solid external masonry walls   AD E: 2.151

No guidance is provided in the approved document. Designers are advised to seek specialist advice.

7.5.4   Junctions with internal framed walls   AD E: 2.152

No restrictions are imposed.

7.5.5   Junctions with internal masonry walls   AD E: 2.153

No restrictions are imposed.

7.5.6   Junctions with internal timber floors   AD E: 2.154

In order to prevent flanking transmission, air paths into the cavity from floor voids must be blocked by solid timber blockings, continuous ring beams or joists.

7.5.7   Junctions with internal concrete floors   AD E: 2.155

No guidance is provided in the approved document. Designers are advised to seek specialist advice.

7.5.8   Junctions with timber ground floors   AD E: 2.156

In order to prevent flanking transmission, air paths into the cavity from floor voids must be blocked by solid timber blockings, continuous ring beams or joists.

7.5.9   Junctions with concrete ground floors   AD E: 2.158

A concrete ground floor slab laid directly on the ground may be of continuous construction under a Type 4 separating wall whereas a suspended concrete ground floor may only pass beneath a Type 4 wall if the floor has a mass per unit area of not less than $365 \, \text{kg/m}^2$.

7.5.10   Junctions with ceilings and roofs   AD E: 2.160–2.163

- The separating wall should preferably continue to the underside of the roof, a flexible closer should be provided at its junction with the roof and the separating wall linings should be sealed with tape or caulked with appropriate sealant at their junction with the ceiling.
- If the roof or loft space is not a habitable room and the ceiling has sealed joints and a mass per unit area of at least $10 \, \text{kg/m}^2$, then within the roof space, either:
  (a)   the lining on each of the two frames may be reduced to two layers of plasterboard, each sheet with a mass per unit area of at least $10 \, \text{kg/m}^2$; or
  (b)   the wall may be reduced to one frame with two layers of plasterboard, each sheet with a mass per unit area of at least $10 \, \text{kg/m}^2$, on both sides of the frame. In this case, the cavity must be closed at ceiling level in such a way that the two frames are not rigidly connected together.
- Cavities in external walls should be closed at their eaves. A flexible material should be used for the closure with no rigid joint between the two leaves. It is important to avoid a rigid connection between the two leaves and if a rigid material is used it should only be rigidly attached to one leaf, see BRE BR 262, *Thermal insulation: avoiding risks*, section 2.3 [17].

Table 7.6   Construction procedures relating to the sound insulation of separating walls.

Separating wall type	Correct procedure	Procedures which must be avoided
1, 2, 3, 4	Control flanking transmission between separating walls and connected walls and floors as described in the text	
	Ensure external cavity walls are stopped with flexible closers at junctions with separating wall as required in text	
1, 2, 3	Fill and seal all masonry joints with mortar	
	Ensure flue blocks will not adversely affect sound insulation. Ensure that a suitable finish is used over flue blocks (see BS 1289-1:1986 and obtain manufacturer's advice)	
1, 2, 4	Stagger the positions of sockets on opposite sides of separating walls. In the case of Type 4 walls, use a similar thickness of cladding material behind socket boxes	In the case of Type 1 and 2 walls, do not use deep sockets and chases in the wall and do not locate them back to back
		In the case of Type 4 walls do not locate sockets back to back or chase plasterboard and a minimum of 150 mm edge to edge stagger is recommended between sockets
1, 2		Do not create a junction between Type 1 and 2 walls in which the cavity is bridged by the solid wall
		Do not attempt to convert a Type 2 separating wall into a Type 1 wall by inserting mortar or concrete into the cavity
1	Lay bricks 'frog up'	
	Use bricks and blocks which extend to the full thickness of the wall	
2	Keep cavity leaves separate below ground floor level	Do not change to a solid wall in the roof space. A rigid connection between the leaves will adversely affect performance
		Do not build a cavity wall off a continuous concrete floor slab
3	Fix panels or supporting frames to ceiling and floor only	Do not fix, tie or otherwise connect the free standing panels to the masonry core wall
	Tape and seal all joints	
4	Ensure that fire stops in the cavity between frames are either flexible or fixed to one frame only	If the two leaves have to be connected for structural purposes, do not use ties of greater cross section than 40 mm by 3 mm. These should be fixed to the studwork at or just below ceiling level. Do not set the ties at closer than 1.2 m centres.
	Ensure that each layer of plasterboard is independently fixed to the frame	

7.5.11   Junctions with separating floors   AD E: 2.164

See section 3 of the AD (Chapter 8 of this book) for details of junctions between separating floors and Type 4 separating walls.

7.6   Walls separating habitable rooms from other parts of a building   AD E: 2.25–2.28

Buildings containing flats and rooms for residential purposes often contain non-habitable spaces juxtaposed to those which are occupied. The approved document stipulates the sound insulation to be provided by the walls of refuse chutes in terms of mass, including finishes, per unit area. If the refuse chute is separated from a habitable room or kitchen its mass per unit area should be $1320\,\text{kg/m}^2$ or more, whereas if it is a non-habitable room the mass per unit area may be reduced to $220\,\text{kg/m}^2$.

Another common source of noise in multi-occupancy buildings is that transmitted from corridors. The AD states that in order to control noise from this source, walls between corridors and flats should be constructed to the standard used for separating walls, see sections 7.2 to 7.5. A weak point in the sound insulation provided from corridor noise is that transmitted through doors. For this reason, the document states that doors to corridors should have good sealing around their perimeters (including, if practical, their thresholds), and a mass per unit area of at least $25\,\text{kg/m}^2$, or a weighted sound reduction index of at least 29dB. The term sound reduction index is described in Chapter 3 and the relevant measurement standards are listed in section 6.9.3.3. In 'noisy' parts of a building, noise should be contained by a lobby, two doors in series (i.e. such that each door is passed through in turn) or a high performance door set, and if this is impossible, flats in the vicinity should be provided with this type of protection at their entrances. It is very important when considering the sound insulation of doors to bear in mind also the requirements of Building Regulations Part B – Fire safety and Part M – Access and facilities for disabled people.

7.7   Construction procedures

The AD provides information regarding correct and incorrect construction procedures which will influence the sound insulation offered by separating walls. This is summarised in Table 7.6.

8 Separating Floors and their Flanking Constructions – New Buildings

Approved Document E, Section 3: Separating floors and associated flanking constructions for new buildings

8.1 Construction forms   AD E: 3.1–3.7 & 3.9

Section 3 of the Approved Document E gives examples of floors which should achieve the performance standards for houses and flats tabulated in section 0: Performance – Table 1a of the AD and in Chapter 5 of this book. However, as is stated in the document, the actual performance of a floor will depend on the quality of its construction. The information in this section of the AD is provided only for guidance; it is not intended to be exhaustive and alternative constructions may achieve the required standard. Designers are also advised to seek advice from other sources such as manufacturers regarding alternative designs, materials or products. The details in no way override the requirement to undergo pre-completion testing. For the provision of floors in 'rooms for residential purposes', see Chapter 11 of this book.

The separating floor constructions described in the AD are divided into three types:

- **Type 1   Concrete base with ceiling and soft floor covering**
- **Type 2   Concrete base with ceiling and floating floor**
  Floors of this type require one of three types of floating floor. These are defined by the terms (a), (b) and (c).
- **Type 3   Timber frame base with ceiling and platform floor**.

The three floor types are shown in Fig. 8.1 and within each floor type, alternatives are presented in a ranking order such that the one which should offer the best sound insulation is described first.

The following notes should be borne in mind when considering the alternatives which are presented.

- The mass per unit area of floors is quoted in kilograms per square metre (kg/m^2) and this may be obtained from manufacturers' data. Alternatively it may be obtained using a procedure described in Annex A of the AD, and outlined in the Appendix to this book. Densities of materials are quoted in kilograms per cubic metre (kg/m^3).
- Acoustically, the mass of a bonded screed acts with the slab onto which it is laid and therefore, where appropriate, may be taken into account when calculating

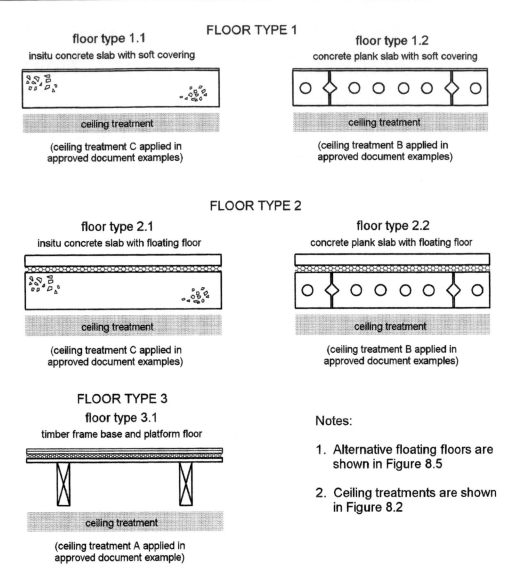

Fig. 8.1 Alternative types of floor construction.

the mass per square metre of a floor, but the mass of a floating screed does not, and therefore must not be taken into account.

- Solid in-situ concrete and hollow plank floors are considered in the guidance notes but beam and block floors are not. It is suggested that designers take advice from manufacturers regarding the sound insulation properties of this latter type of floor.

8.1.1 Floor penetrations   AD E: 3.41–3.43, 3.79–3.82 & 3.117–3.120

Particular attention must be given to pipes or ducts which pass through separating floors. Points made in the approved document include the following:

- Pipes and ducts penetrating a floor separating habitable rooms in different dwelling places should be enclosed for their full height, i.e. for the full height of the room if they pass from floor to ceiling, with material which has a mass per unit area of at least $15 \, \text{kg/m}^2$ and either: the pipe or duct should be wrapped, or the enclosure lined, with 25 mm of unfaced mineral wool.
- If the floor incorporates a floating layer, a gap of approximately 5 mm should be left between this and the enclosure with the gap filled with sealant or neoprene. The enclosure may extend down to the floor base but if so it must be isolated from the floating layer.
- There are fire safety implications where separating floors are penetrated, and fire protection to satisfy the requirements of Building Regulation Part B must be provided. Flexible fire stopping should be used and, in order to prevent structure-borne transmission, it is important that there is no rigid contact between pipes and floor.
- Ducts containing gas pipes must be ventilated at each floor. Gas pipes may be housed in separate ventilated ducts or, alternatively, gas pipes need not be enclosed. It is essential that relevant codes and standards are complied with to ensure gas safety. Reference: The Gas Safety (Installation and Use) Regulations 1998, SI 1998 No. 2451

8.1.2 Ceiling construction   AD E: 3.17–3.22

The type of ceiling provided has a significant bearing on the acoustic performance of a floor and so, since alternative ceiling treatments may be combined with each type of floor, the three ceiling treatments described in the AD are defined by the letter A, B or C and each floor type is qualified by the letter A, B or C depending on its ceiling treatment. The AD states that the ceiling treatments, as described in Table 8.1 and shown in Fig. 8.2, are ranked such that A should provide the highest level of sound insulation and, further, that if a ceiling treatment of a higher ranking than one applied in a guidance example is used, then providing that there is no significant flanking transmission, the result should be improved sound insulation from the complete floor.

8.1.3 Junctions and flanking transmission   AD E: 3.10

It is very important that the junctions between a separating floor and the elements to which it is attached are carefully designed because this will have a significant influence on its overall effectiveness. Since the level of sound insulation achieved will be strongly dependent on the extent to which flanking transmission exists,

Table 8.1 Ceiling treatments for separating floor constructions.

Ceiling treatment	Form	Specification
A	Independent ceiling with absorbent material	At least two layers of plasterboard with staggered joints and total mass per unit area of at least 20 kg/m^2 Mineral wool laid in cavity above ceiling to provide sound absorption. Minimum thickness 100 mm, minimum density 10 kg/m^3 See note below regarding fixings
B	Plasterboard on proprietary resilient bars with absorbent material to fill ceiling void	Single layer of plasterboard with mass per unit area of at least 10 kg/m^2 Plasterboard fixed using proprietary resilient bar. For concrete floor, fix bar to timber battens. Bar should be fixed in accordance with manufacturer's instructions Fill ceiling void with mineral wool with density of at least 10 kg/m^3
C	Plasterboard on timber battens, or plasterboard on resilient channels with absorbent material to fill ceiling void	Single layer of plasterboard with mass per unit area of at least 10 kg/m^2 Plasterboard should be fixed using proprietary resilient channels or timber battens Fill ceiling void with mineral wool with density of at least 10 kg/m^3 if resilient channels are used

Notes:
(a) If using ceiling treatment A with floor types 1 or 2, the ceiling should be attached to independent joists supported only by surrounding walls with a clearance of a minimum of 100 mm between top of ceiling plasterboard and underside of base floor. If using floor type 3, the ceiling should be attached to independent joists supported by surrounding walls with extra support from resilient hangers attached directly to the floor. In this case there should be a clearance of a minimum of 100 mm between the top of the ceiling joists and underside of base floor.
(b) Light fittings recessed into ceilings may reduce sound insulation.
(c) See BRE BR 262 *Thermal Insulation: avoiding risks*, section 2.4 [17] regarding heat emission from electrical cables which may be covered by absorbent material.

considerable advice is given in the AD on this issue. Guidance is provided on the junctions between separating floors and the following other elements:

- External cavity walls with masonry inner leaf
- External cavity wall with timber framed inner leaf
- Solid masonry external walls
- Internal walls – masonry
- Internal walls – timber
- Floor penetrations (see section 8.1.1)
- Separating walls Type 1, 2, 3 and 4 (these being relevant to the design of flats).

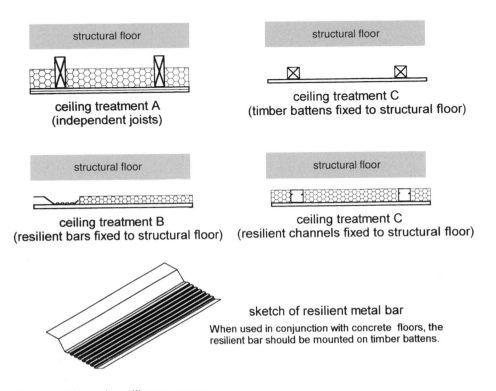

Fig. 8.2 Alternative ceiling treatments.

However, not all floor/junction combinations are considered. It is stated in the AD that if any element has a separating function, e.g. a ground floor which is also a separating floor for a basement flat, then the separating element requirements should take precedence.

8.2 Design of floor Type 1 – concrete base with ceiling and soft floor covering   AD E: 3.23–3.30

It is the large mass of the concrete slab together with that of the ceiling which provides airborne sound insulation from this type of floor. For impact sound insulation it relies on the soft covering preventing the contact of hard surfaces.

To be acceptable, a soft covering should be a resilient material, or have a resilient base, with a thickness, when uncompressed, of 4.5 mm or more. An alternative approach is to measure the 'weighted reduction of impact sound pressure level', ΔL_w, of the floor covering in accordance with the procedure described in Appendix B3 of the approved document, see section 6.9.3.1 of this book (Floor coverings and floating floors). To be acceptable the measured value of ΔL_w should be 17dB or more.

Two variations of floor Type 1 are described in Approved Document E, which according to the document should, if built correctly, comply with Requirement E1, see Table 8.2. The first floor has ceiling treatment C and hence is referenced 1.1C, whereas the second has ceiling treatment B and hence is referenced 1.2B.

Table 8.2   Examples of concrete base with ceiling and soft covering floor constructions.

	Floor Type	
	1.1C	**1.2B**
Description	Solid concrete slab, cast in-situ with or without permanent shuttering. Soft floor covering and ceiling treatment C	Hollow or solid concrete planks. Soft floor covering and ceiling treatment B
Minimum mass per unit area of concrete base	365 kg/m^2 including any bonded screed. Permanent shuttering of solid concrete or metal may be also be included	365 kg/m^2 including any bonded screed
Floor covering	Soft floor covering. This is essential	Soft floor covering. This is essential
Ceiling treatment	C or better is essential	B or better is essential
Other requirements		Use regulating floor screed
		All joints between and around planks to be fully grouted to ensure complete air tightness

In order for the performance of Type 1 floors to be satisfactory, the junctions between them and their surrounding elements of constructions must be designed correctly. The requirements detailed in the approved document are explained in the following sections. See also section 8.5 for information relating to correct and incorrect construction procedures.

8.2.1   Junctions with external cavity walls which have masonry inner leaves AD E: 3.31–3.35

- There are no restrictions on the outer leaf of the wall.
- Unless the cavity is fully filled with mineral wool, expanded polystyrene beads or some other suitable insulating material (the AD states that manufacturers' advice should be sought regarding alternative suitable materials), the cavity should be stopped with a flexible closer as shown in Fig. 8.3. The flexible closer must be protected from the effects of moisture and it is essential that adequate drainage is provided; note the provision of a flexible cavity tray, or an equivalent, as shown in Fig. 8.3.
- The mass per unit area of the inner leaf of the external wall should be at least 120 kg/m^2 excluding its finish in order to reduce the effects of flanking transmission.

- The floor base, but not its screed, should continue to the face of the cavity without bridging the cavity and if floor Type 1.2B is used, with the planks laid parallel to the wall, the first joint between planks should be 300 mm or more from the cavity face, as shown in Fig. 8.3.
- The use of wall ties in external masonry walls is considered in section 2 of the approved document. See section 7.1.1 of this book.

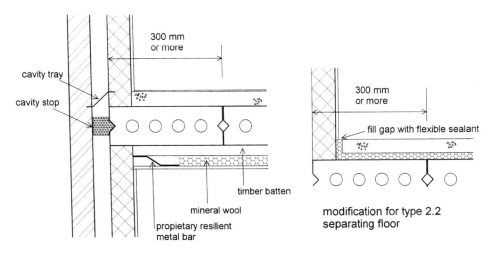

type 1.2B separating floor

Fig. 8.3   Details of junctions between Type 1 and 2 separating floors and external cavity walls.

8.2.2   Junctions with external cavity walls which have timber framed inner leaves   AD E: 3.36

- There are no restrictions on the outer leaf of the wall.
- It is assumed that the cavity will not be filled and so the cavity should be stopped with a flexible closer.

The finish of the inner leaf should consist of two layers of plasterboard each with a mass per unit area of at least $10 \, \text{kg/m}^2$ with all joints sealed with tape or caulked with appropriate sealant.

8.2.3   Junctions with solid external masonry walls   AD E: 3.37

No guidance is provided in the AD. Designers are advised to seek specialist advice.

8.2.4   Junctions with internal framed walls   AD E: 3.38

No restrictions are imposed.

8.2.5   Junctions with internal masonry walls   AD E: 3.39–3.40

The floor base should be continuous through or above such walls and any internal load-bearing wall or any internal wall rigidly connected to a separating floor should have a mass per unit area of at least $120\,kg/m^2$ excluding finish.

8.2.6   Junctions with solid masonry (Type 1) separating walls
AD E: 3.44–3.45

Floors bases of Type 1.1C should pass through Type 1 separating walls. Floor bases of Type 1.2B should not be continuous through Type 1 separating walls. In neither case should any screed penetrate the separating wall. See Fig. 8.4 for examples and requirements of both types of construction.

8.2.7   Junctions with cavity masonry (Type 2) separating walls
AD E: 3.46–3.48

- The mass per unit area of any leaf, excluding finish, that supports or adjoins the floor should be $120\,kg/m^2$ or more.
- The floor base, but not its screed, should continue to the face of the cavity without bridging the cavity and if floor Type 1.2B is used, with the planks laid parallel to the wall, the first joint between planks should be 300 mm or more from the cavity face as shown in Fig. 8.4.

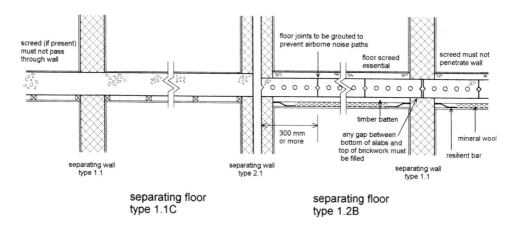

Fig. 8.4   Details of junctions between Type 1 separating floors and Type 1 and 2 separating walls.

8.2.8   Junctions with masonry between independent panels (Type 3) separating walls   AD E: 3.49–3.54

Assuming the separating wall has a solid core, i.e. wall Types 3.1 and 3.2:

- Floor bases of Type 1.1C should pass through the wall. Floor bases of Type 1.2B should not be continuous through this type of separating wall. In neither case should any screed penetrate the separating wall. Construction should be similar to that shown in Fig. 8.4 for solid masonry separating walls.
- If floor Type 1.2B is used in conjunction with wall Type 3.2, with the planks laid parallel to the wall, the first joint between planks should be 300 mm or more from the centreline of the masonry core.

Assuming the separating wall has a cavity core, i.e. wall Type 3.3:

- The mass per unit area of any leaf that is supporting or adjoining the floor, excluding finish, should be 120 kg/m^2 or more.
- The floor base, but not its screed should continue to the face of the cavity without bridging the cavity and if floor Type 1.2B is used, with the planks laid parallel to the wall, the first joint between planks should be 300 mm or more from the cavity face of the adjacent leaf of the masonry core. Construction should be similar to that shown in Fig. 8.4 for cavity masonry separating walls.

8.2.9   Junctions with framed with absorbent material (Type 4) separating walls   AD E: 3.55

No guidance is provided in the AD. Designers are advised to seek specialist advice.

8.3   Design of floor Type 2 – concrete base with ceiling and floating floor   AD E: 3.56–3.60

The complete floor construction consists of a concrete floor base on top of which is a resilient layer. On top of the resilient layer there is a solid 'floating' layer and below the concrete floor base is a ceiling which, as has previously been defined, may be classified as Type A, B or C. The floating floor consists of the floating layer and the resilient layer.

The sound insulation offered by floors of this type depends on the mass of the concrete base, the floating layer and the ceiling. It is also improved by the isolation of the floating layer and the ceiling. Impact sound is reduced at source by the floating layer. However, even if resistance to airborne sound only is required, the full construction should still be used.

8.3.1   Floating floors   AD E: 3.62–3.66

Examples of three types of floating floors are presented in the AD. These are described in Table 8.3 and shown in Fig. 8.5.

Table 8.3   Alternative types of floating floors as described in the approved document.

	Floating Floor Type		
	(a)	**(b)**	**(c)**
Description	Timber raft floating layer supported on a resilient layer	Sand and cement screed floating layer supported on a resilient layer	A floating floor designed to satisfy a performance criterion
Floating layer	Raft of board material with bonded edges, such as t. & g. timber with mass per unit area of not less than 12 kg/m^2, fixed to 45 mm × 45 mm battens Raft laid on resilient layer without any fixings. Do not lay battens along joints in resilient layer	65 mm of sand and cement or suitable proprietary screed product Screed to have mass per unit area of 80 kg/m^2 or more Resilient layer must be protected while screed is laid, e.g. by using a 20–50 mm wire mesh	The floating floor should consist of a rigid board above a resilient and/or damping layer which provides a weighted reduction in impact sound pressure level, ΔL_{w}, of 29 dB or more
Resilient layer	Mineral wool with density of 36 kg/m^3 and thickness of at least 25 mm. The layer of mineral wool may have a paper faced underside	Mineral wool with density of 36 kg/m^3 and thickness of at least 25 mm. The layer of mineral wool to be paper faced on upper side to prevent wet screed entering the resilient layer, or, a layer meeting the specific dynamic stiffness criterion of 15 MN/m^3 and min thickness of 5mm under specified load. See BS 29052-1:1992	ΔL_{w} is the improvement in impact sound insulation obtained in a laboratory by installing a floating floor over a test floor Laboratory measurement should be in accordance with the procedure described in section 6.9.3.1

Designers are advised to take advice from manufacturers on proprietary screed products and the performance and installation of proprietary floating floors.

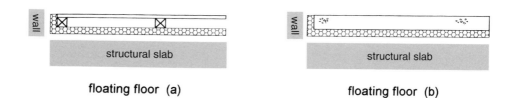

floating floor (a)   floating floor (b)

Fig. 8.5   Alternative floating floor constructions.

8.3.2   Floor, floating floor, ceiling combinations   AD E: 3.67–3.68

Two variations of floor Type 2 are described in Approved Document E, which according to the document should, if built correctly, comply with Requirement E1, see Table 8.4. The first floor has ceiling treatment C and hence is referenced 2.1C, whereas the second has ceiling treatment B and hence is referenced 2.2B.

When considering the alternatives in Table 8.4, it should be borne in mind that in the approved document, alternatives are ranked such that the one which should give the best performance appears first. Also, the ceiling treatment specified is that which is required to meet the requirements of the AD. Since, as already stated, the ceilings are ranked in descending order of performance from A to C, substitution of Type A or B ceilings in Floor 2.1 or a Type A ceiling in Floor 2.2 should give improved sound insulation.

In order for the performance of Type 2 floors to be satisfactory, the junctions between them and their surrounding elements of constructions must be designed correctly. The requirements detailed in the approved document are explained in the following sections. See also section 8.5 for information relating to correct and incorrect construction procedures.

Table 8.4   Examples of concrete base with ceiling and floating floor constructions.

	Floor Type	
	2.1C	**2.2B**
Description	Floating floor	Floating floor
	Solid concrete slab, cast in situ with or without permanent shuttering	Hollow or solid concrete planks
	Ceiling treatment C	Ceiling treatment B
Minimum mass per unit area of concrete base	300 kg/m^2 including any bonded screed. Permanent shuttering of solid concrete or metal may also be included	300 kg/m^2 including any bonded screed
Floor covering	Floating floor (a), (b) or (c) is essential	Floating floor (a), (b) or (c) is essential
Ceiling treatment	C or better is essential	B or better is essential
Other requirements	Regulating floor screed is optional	Use regulating floor screed
		All joints between and around planks to be fully grouted to ensure complete air tightness

8.3.3 Junctions with external cavity walls which have masonry inner leaves AD E: 3.69–3.73

- There are no restrictions on the outer leaf of the wall.
- Unless the cavity is fully filled with mineral wool, expanded polystyrene beads or some other suitable insulating material (the AD states that manufacturers' advice should be sought regarding alternative suitable materials), the cavity should be stopped with a flexible closer as shown in Fig. 8.3. The flexible closer must be protected from the effects of moisture and it is essential that adequate drainage is provided; note the provision of a flexible cavity tray, or an equivalent, as shown in Fig. 8.3.
- The mass per unit area of the inner leaf of the external wall should be at least 120 kg/m^2 excluding its finish in order to reduce the effects of flanking transmission.
- The floor base, but not its screed, should continue to the face of the cavity without bridging the cavity and if floor Type 2.2B is used, with the planks laid parallel to the wall, the first joint between planks should be 300 mm or more from the cavity face, as shown in Fig. 8.3.
- Treatment of the intersection of the floating floor and the wall should be as shown in Figs. 8.3 and 8.6.
- The use of wall ties in external masonry walls is considered in section 2 of the approved document. See section 7.1.1 of this book.

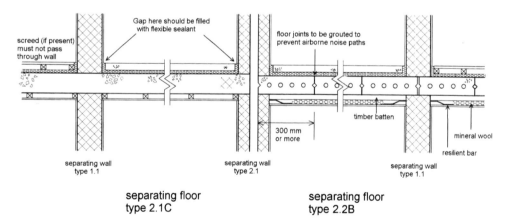

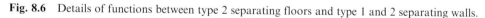

Fig. 8.6 Details of functions between type 2 separating floors and type 1 and 2 separating walls.

8.3.4 Junctions with external cavity walls which have timber framed inner leaves   AD E: 3.74

- There are no restrictions on the outer leaf of the wall.
- It is assumed that the cavity will not be filled and so the cavity should be stopped with a flexible closer.

- The finish of the inner leaf should consist of two layers of plasterboard each with a mass per unit area of at least $10 \, \text{kg/m}^2$ with all joints sealed with tape or caulked with appropriate sealant.

8.3.5   Junctions with solid external masonry walls   AD E: 3.75

No guidance is provided in the AD. Designers are advised to seek specialist advice.

8.3.6   Junctions with internal framed walls   AD E: 3.76

No restrictions are imposed.

8.3.7   Junctions with internal masonry walls   AD E: 3.77–3.78

The floor base should be continuous through or above such walls and any internal load bearing wall or any internal wall rigidly connected to a separating floor should have a mass per unit area of at least $120 \, \text{kg/m}^2$ excluding finish.

8.3.8   Junctions with solid masonry (Type 1) separating walls
AD E: 3.83–3.84

Floors bases of Type 2.1C should pass through Type 1 separating walls. Floor bases of Type 2.2B should not be continuous through Type 1 separating walls. In neither case should any screed penetrate the separating wall. See Fig. 8.6 for examples and requirements of both types of construction.

8.3.9   Junctions with cavity masonry (Type 2) separating walls
AD E: 3.85–3.86

The floor base, but not its screed, should continue to the face of the cavity without bridging the cavity and if floor Type 2.2B is used, with the planks laid parallel to the wall, the first joint between planks should be 300 mm or more from the cavity face as shown in Fig. 8.6.

8.3.10   Junctions with masonry between independent panels (Type 3)
separating walls   AD E: 3.87–3.92

Assuming the separating wall has a solid core, i.e. wall Types 3.1 and 3.2:

- Floor bases of Type 2.1C should pass through the wall. Floor bases of Type 2.2B should not be continuous through this type of separating wall. In neither case should any screed penetrate the separating wall. Construction should be as shown in Fig. 8.7.

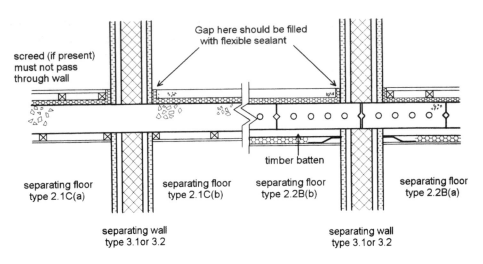

Fig. 8.7   Details of junctions between Type 2 separating floors and Type 3 separating walls.

- If floor Type 2.2B is used in conjunction with wall Type 3.2, with the planks laid parallel to the wall, the first joint between planks should be 300 mm or more from the centreline of the masonry core.

Assuming the separating wall has a cavity core, i.e. wall Type 3.3:

- The mass per unit area of any leaf that is supporting or adjoining the floor, excluding finish, should be 120 kg/m^2 or more.
- The floor base, but not its screed, should continue to the face of the cavity without bridging the cavity and if floor Type 2.2B is used, with the planks laid parallel to the wall, the first joint between planks should be 300 mm or more from the cavity face of the adjacent leaf of the masonry core.

8.3.11   Junctions with framed with absorbent material (Type 4) separating walls   AD E: 3.93

No guidance is provided in the AD. Designers are advised to seek specialist advice.

8.4   Design of floor Type 3 – timber frame base and platform floor with ceiling treatment   AD E: 3.94–3.102

Floors of this type consist of a structural timber base comprising boarding supported on timber joists upon which is a floating layer resting on a resilient layer. The platform floor consists of the floating layer and the resilient layer. Beneath the structural floor is a Type A ceiling treatment. The construction form is shown in Fig.

8.8. In order to obtain good insulation against both airborne and impact sound transmission it is essential that there is good acoustic isolation between the platform floor and the structural base, and between the structural base and the ceiling. The platform floor is important because it reduces the effects of impact noise at source. However, even if resistance to airborne sound only is required, the full construction should still be used.

There are fewer variations of this floor than there are of floor Types 1 and 2 and only one construction form is described in the AD. According to the document this should, if built correctly, comply with Requirement E1. Since this example, which is described in Table 8.5 and shown in Fig. 8.8, utilises ceiling treatment A it is referred to as separating floor Type 3.1A.

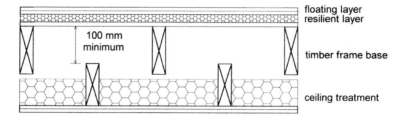

Fig. 8.8 Construction form of floor type 3.1A.

In order for the performance of Type 3 floors to be satisfactory, the junctions between them and their surrounding elements of construction must be designed correctly. The requirements detailed in the approved document are explained in the following sections. See also section 8.5 for information relating to correct and incorrect construction procedures.

8.4.1   Junctions with external cavity walls which have masonry inner leaves AD E: 3.103–3.108

- There are no restrictions on the outer leaf of the wall.
- Unless the cavity is fully filled with mineral wool, expanded polystyrene beads or some other suitable insulating material (the approved document states that manufacturers' advice should be sought regarding alternative suitable materials), the cavity should be stopped with a flexible closer.
- The AD states that it is necessary to line the internal face of the masonry inner leaf of the cavity wall with independent panels as is described for Type 3 separating walls. However, if the mass per unit area of the inner leaf is greater than $375 \, \text{kg/m}^2$, the independent panels need not be provided.
- The ceiling, which consists of independent joists supporting plasterboard, must continue to the masonry of the inner leaf. However, where the ceiling passes above the independent panels it should be sealed with tape or caulked with appropriate sealant.

Table 8.5   Example of timber frame base with ceiling and platform floors.

Floor Type 3.1A	
Description	Timber joists supporting a deck and platform floor (consisting of resilient layer and floating layer) together with ceiling treatment A beneath the structural floor
Structural floor	Timber joists selected to satisfy structural requirements together with a deck which has a minimum mass per unit area of 20 kg/m^2
Floating layer	A minimum of two layers of board each with a minimum thickness of 8 mm to provide a total mass per unit area of at least 25 kg/m^2. The layers should be fixed together ensuring staggered joints and laid loose on the resilient layer
	Two example floating layer constructions are provided in Approved Document E. These are as follows:
	1. 18 mm of timber or wood based board with tongued and grooved edges and glued joints. These should be spot bonded to a substrate of 19 mm thick plasterboard with staggered joints to give a total mass per unit area of at least 25 kg/m^2
	2. Two layers of cement bonded particle board glued and screwed together with staggered joints. The resulting platform must have a total thickness of 24 mm and a mass per unit area of at least 25 kg/m^2
Resilient layer	Mineral wool, which may be paper faced on its underside, with a thickness of at least 25 mm and density in the range 60 to 100 kg/m^3
	If the material chosen is towards the lower end of the above range, impact sound insulation is improved but it may result in what is described in the document as a 'soft' floor, i.e. one which may be considered to respond excessively to fluctuating loads. It is suggested in the document that this may be overcome by providing additional support via a timber batten which is fixed to the walls and has a foam strip along its top
Ceiling treatment	Ceiling treatment A must be used in conjunction with this floor.

- The method of connecting the floor base, i.e. the floor joists, to the external wall is not prescribed but it is emphasised that there must be no air paths connecting the floor to the wall cavity. This would pose a source of flanking transmission which it would be difficult to remedy at a later date.
- The use of wall ties in external masonry walls is considered in section 2 of the AD. See section 7.1.1 of this book.

8.4.2   Junctions with external cavity walls which have timber framed inner leaves   AD E: 3.109–3.112

- There are no restrictions on the outer leaf of the wall.
- It is assumed that the cavity will not be filled and so the cavity should be stopped with a flexible closer.

- The finish of the inner leaf should consist of two layers of plasterboard each with a mass per unit area of at least $10\,kg/m^2$ with all joints sealed with tape or caulked with appropriate sealant.
- The method of connecting the floor base, i.e. the floor joists, to the external wall is not prescribed but where floor joists are perpendicular to the wall, the spaces between the floor joists should be sealed to the full depth of the floor with timber blocking.
- The junction between the ceiling and wall lining should be sealed with tape or caulked with appropriate sealant.

8.4.3   Junctions with solid external masonry walls   AD E: 3.113

No guidance is provided in the AD. Designers are advised to seek specialist advice.

8.4.4   Junctions with internal framed walls   AD E: 3.114–3.115

- Where floor joists are perpendicular to the wall, the spaces between the floor joists should be sealed to the full depth of the joists with timber blocking.
- The junction between the ceiling and the wall should be sealed with tape or caulked with appropriate sealant.

8.4.5   Junctions with internal masonry walls   AD E: 3.116

No guidance is provided in the AD. Designers are advised to seek specialist advice.

8.4.6   Junctions with solid masonry (Type 1) separating walls
AD E: 3.121–3.122

- When floor joists are to be supported on this type of wall, they should not be built into the wall but must be supported on joist hangers.
- The junction between the ceiling and the wall should be sealed with tape or caulked with appropriate sealant.

8.4.7   Junctions with cavity masonry (Type 2) separating walls
AD E: 3.123–3.126

- When floor joists are to be supported on this type of wall, they should not be built into the wall but must be supported on joist hangers.
- The AD states that it is necessary to line the leaf of the cavity wall which is nearest to the room in question with independent panels as is described for Type 3 separating walls. However, if the mass per unit area of the nearest leaf to the room in question is greater than $375\,kg/m^2$, the independent panels need not be provided.
- The ceiling, which consists of independent joists supporting plasterboard, should continue to the masonry of the wall. However, where the ceiling passes above the

Table 8.6 Construction procedures relating to the sound insulation of separating floors.

Element	Correct procedure	Procedures which must be avoided
Floor Types 1, 2 and 3	Seal the perimeter of independent ceilings with tape or sealant	Do not create a rigid or direct connection between an independent ceiling and its floor base
	Give extra attention to workmanship and detailing at perimeter and where there are floor penetrations, to reduce flanking transmission and prevent air paths	
	Ensure notes on junction construction are complied with to reduce flanking effects	
Floor Types 1 and 2	Fill all joints between floor components to avoid air paths between floors	Do not allow a floor base to bridge across a cavity in a masonry cavity wall
	Build concrete separating floors (or floor bases) into all of their masonry perimeter walls	
	Ensure that all gaps between heads of masonry walls and undersides of concrete floors are filled with mortar	
Floor Type 1	Fix or glue the soft covering to the floor	A soft floor covering must be used. Do not use ceramic floor tiles, wood blocks or any other non-resilient floor finish rigidly connected to the floor base
Floor Type 2	Leave small gap of size recommended by manufacturers between floating layer and wall at room edge and fill with flexible sealant	Do not bridge between the floating layer and the base or surrounding walls, e.g. with services or fixings which penetrate the resilient layer between floating layer and floor base or walls
	Leave gap of about 5 mm between skirting and floating layer and fill with flexible sealant	Do not allow the floating layer to bridge to the floor base or walls, e.g. through a void in the resilient layer
	Lay rolls or sheets of resilient materials with lapped joints or tightly butted and taped joints	
	Prevent screed entering resilient layers by laying fibrous materials with paper facing uppermost	
Floor Type 3	With respect to the platform floor:	Do not bridge between the floating layer and the timber frame base or surrounding walls, e.g. with services or fixings which penetrate the resilient layer between the floating layer and timber base or walls
	(a) Use the correct density of resilient layer and ensure that it is adequate to carry the applied load	
	(b) Use a resilient material, e.g. expanded or extruded polystyrene strip around the perimeter to ensure that a gap is maintained between wall and floating layer during construction. The strip should be about 4 mm higher than top surface of the floating layer. The gap may be filled with a flexible sealant	
	(c) Lay sheets of resilient materials with tightly butted and taped joints	

independent panels it should be sealed with tape or caulked with appropriate sealant.

8.4.8   Junctions with masonry between independent panels (Type 3) separating walls   AD E: 3.127–3.128

- When floor joists are to be supported on this type of wall, they should not be built into the wall but must be supported on joist hangers.
- The ceiling, which is supported on independent joists, should pass through the independent panels to the masonry core. The junction where the ceiling passes over the independent panels should be sealed with tape or caulked with appropriate sealant.

8.4.9   Junctions with framed with absorbent material (Type 4) separating walls   AD E: 3.129–3.130

- Where floor joists are perpendicular to the wall, the spaces between the floor joists should be sealed to the full depth of the floor with timber blocking.
- The junction between the ceiling and the wall lining should be sealed with tape or caulked with appropriate sealant.

8.5   Construction procedures

The AD provides information regarding correct and incorrect construction procedures which will influence the sound insulation offered by separating floors. This is summarised in Table 8.6.

9 Dwelling Houses and Flats Formed by Material Change of Use

Approved Document E, Section 4: Dwelling-houses and flats formed by material change of use

9.1 General requirements   AD E: 4.1–4.21

When converting a building to dwelling places, consideration must be given to satisfying the requirements of Part E. The complexities of sound transmission in a building that is subject to conversion are considerable, and it must be noted that a superficial or incomplete approach to possible solutions is fraught with danger. For this reason, high quality acoustic advice is almost always essential at an early stage of the design.

Section 4 of Approved Document E describes forms of treatment to walls, floors, stairs and their associated junctions which it may be appropriate to apply when converting existing buildings by material change of use to provide dwelling houses and flats. Rooms for residential purposes are considered in section 6 of the AD, see Chapter 11 of this book. The point is made in the AD that existing components may already satisfy the requirements of Table 1a of the document, which are outlined in Chapter 5, and it is suggested that this would be so if the construction of a component, its junction detailing and flanking construction was sufficiently similar to that of one of the example walls or floors presented for new buildings. It also states that in some circumstances the example separating walls and floors (including their flanking constructions) provided for the construction of new buildings in sections 2 and 3 of the AD may be used to provide guidance as to what work may be necessary to bring existing components into line with the required standard. The importance of the control of flanking transmission, see section 9.6, in instances of change of use is also emphasised in the AD.

The construction of an existing component may be such that a designer has little idea as to whether it will comply with the requirements of Table 1a of the AD. To help in this case, treatments to existing components are provided in the AD as follows:

- Walls, one form of treatment
- Floors, two forms of treatments
- Stairs, one form of treatment.

The nature of each of these is explained below. It is important to note that the AD states that the example treatments 'can be used to increase sound insulation', but not that if constructed correctly, treated components will achieve the required standard. This is because the level of insulation reached will depend heavily on the existing construction. It is also stated in the AD that the information provided is only guidance and that other designs, materials or products may be used to achieve the required performance standard. Designers are recommended to seek advice from manufacturers and/or other expert sources.

It is suggested in the AD that, with respect to requirements for mass, for example, an existing component may be sufficiently similar to comply with the required performance requirements if its mass per unit area is within 15% of that of an equivalent component recommended for the construction of new buildings. This may be the case, but due to uncertainty about the workmanship and density of material used throughout the component, additional treatment may still be advisable.

In view of the complex nature of the construction forms resulting from conversion work, it may be necessary to seek specialist acoustic advice. The AD cites, as an example of a construction which will require special attention, the consequences of constructing a wall or floor across an existing continuous floor or wall such that the original floor or wall becomes a flanking element. The nature of the building may make the simple isolation of such flanking components technically difficult and in such circumstances and when significant additional loads are imposed on the building as a result of floor and wall treatment, structural advice should be sought.

The AD recommends that the following work should be undertaken to existing floors before the recommended treatments to improve sound insulation are applied:

- Since gaps in existing floorboards are a potential source of airborne sound transmission they should be covered with hardboard or filled with sealant.
- Replacement of floorboards. If required, these should be at least 12 mm thick and mineral wool with a density of not less than $10\,kg/m^2$ should be laid between joists to a depth of not less than 100 mm.
- The mass per unit area of concrete floors should be increased to at least $300\,kg/m^2$, air gaps within them should be sealed and a regulating screed provided if required.
- Existing lath and plaster ceilings should be retained as long as they comply with Building Regulation B – Fire Safety, whereas other ceilings should be modified such that they consist of at least two layers of plasterboard with staggered joints and a total mass per unit area of $20\,kg/m^2$.

Sound transmission from corridors and, particularly, through doors in corridors is often a serious source of annoyance in dwellings formed by material change of use. It is suggested in the AD that the separating walls described in this chapter, i.e. section 4 of the AD, should be used to control sound insulation and flanking transmission between houses or flats formed by material change of use and corridors. The AD states that corridor doors and doors to noisy parts of a building should be constructed as the equivalent doors for new buildings. These are considered in section 7.6 of this book.

9.2   Wall treatment 1: independent panel(s) with absorbent material   AD E: 4.22–4.24

This treatment consists of adding a panel, or panels, to an existing wall which has poor sound insulating properties and the result may be similar to the 'masonry wall with independent panels' as described for new constructions but with insulation in the void between the two. The existing wall will offer some sound insulation and this will be enhanced by the provision of an independent panel, the isolation of that panel and the provision of absorbent material between the two.

The AD suggests that if the original wall is of masonry construction, is 100 mm or more thick and is plastered on both sides, one independent panel will suffice, but in the case of different types of existing wall, a panel should be built on each side of the wall.

The specification of Wall Treatment 1 is as described in Table 9.1 and shown in Fig. 9.1. See also section 9.7 for information relating to correct and incorrect construction procedures.

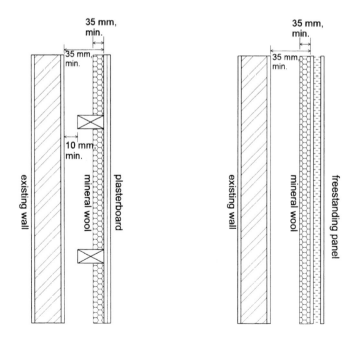

Fig. 9.1   Construction form of Wall Treatment 1.

Table 9.1   Construction form of Wall Treatment 1.

	Description
Construction form	Plasterboard supported on a timber frame or freestanding panels consisting of plasterboard sandwiching a cellular core
Panels	The mass per unit area of each panel should be at least 20 kg/m^2 excluding the mass of any framework
	If supported on a frame, panels should consist of at least two layers of plasterboard with staggered joints. Alternatively, freestanding panels, e.g. two sheets of plasterboard separated by a cellular core, may be used.
Spacing	If panels are supported on a frame there should be a gap of at least 10 mm between the frame and the existing wall. If freestanding panels are used, there should be a gap of at least 35 mm between them and the existing wall.
Mineral wool	Mineral wool with a minimum density of 10 kg/m^2 and minimum thickness of 35 mm to be located in the cavity formed between the panel and the existing wall.

9.3   Floor treatment 1: independent ceiling with absorbent material   AD E: 4.26–4.29

This treatment consists of adding an additional independent ceiling comprising plasterboard, joists and absorbent infill to an existing conventional floor comprising joists, boarding and plasterboard. The mass of the new ceiling, its absorbent infill together with its acoustic isolation will add significantly to the airborne and impact sound insulation offered by the original ceiling. Further, for good insulation, the construction should be made as airtight as possible.

The specification of Floor Treatment 1 is as described in Table 9.2. See also section 9.7 for information relating to correct and incorrect construction procedures.

The point is made in the AD that adoption of this procedure significantly reduces the floor to ceiling height in the converted building and that this should be borne in mind at design stage, i.e. the reduction will be a minimum of 125 mm plus the thickness of the plasterboard and even more if the joists need to be deeper than 100 mm. This can present a problem if window heads are close to the height of the original ceiling and a method of solution is proposed. This, together with the wall junction treatment that should be used when this procedure is employed, is shown in Fig. 9.2.

9.4   Floor treatment 2: platform floor with absorbent material AD E: 4.31–4.33

This treatment consists of adding a platform floor, i.e. a floating layer supported on a resilient layer, to an existing conventional floor comprising timber joists, boarding

Table 9.2   Construction form of Floor Treatment 1.

	Description
Construction form	Independent joist and plasterboard ceiling with absorbent mineral wool infill constructed below a conventional timber joist floor
Independent ceiling	Two or more layers of plasterboard having staggered joints and a mass per unit area of at least 20 kg/m^2
Mineral wool	Mineral wool with a minimum density of 10 kg/m^3 and minimum thickness of 100 mm to be located between the joists in the void between the new and old ceilings
Ceiling support	Independent joists fixed to surrounding walls
	There should be either:
	(a) no further support with a clearance of 25 mm or more between top of the joists and underside of the existing floor; or
	(b) additional support by resilient hangers fixed directly to the underside of the existing floor base
Existing ceiling	Upgrade as necessary to provide two or more layers of plasterboard having staggered joints and a mass per unit area of at least 20 kg/m^2

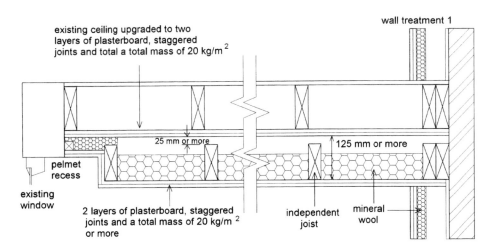

Fig. 9.2   Construction details of Floor Treatment 1 showing junction with Wall Treatment 1.

and plasterboard ceiling. The increase in the total mass of the floor, the action of the resilient layer and the absorbent material should all serve to increase the airborne and impact sound resistance of the floor.

The specification of Floor Treatment 2 should be as described in Table 9.3 and the construction form together with a junction treatment recommended in the AD are shown in Fig. 9.3. See also section 9.7 for information relating to correct and incorrect construction procedures.

Table 9.3   Construction form of Floor Treatment 2.

	Description
Construction form	A floating layer laid on a resilient layer which has first been laid onto the boarding of a conventional timber joist floor
Floating layer	A minimum of two layers of board each with a minimum thickness of 8 mm to provide a total mass per unit area of at least 25 kg/m². The layers should be fixed together, for example by spot bonding or gluing and screwing, ensuring staggered joints and then laid loose on the resilient layer
	Two examples of how a platform floor/floating layer of this type could be constructed are provided in the approved document, see new floor Type 3.1A
Resilient layer	Mineral wool, which may be paper faced on its underside, with a thickness of at least 25 mm and density in the range 60 to 100 kg/m³
	If the material chosen is towards the lower end of the above range, sound insulation is improved but it may result in what is described in the document as a 'soft' floor, i.e. one which may be considered to respond excessively to fluctuating loads. It is suggested in the document that this may be overcome by providing additional support via a timber batten which is fixed to the walls and has a foam strip along its top
Mineral wool	Mineral wool with a minimum density of 10 kg/m³ and minimum thickness of 100 mm to be located between the joists in the floor cavity
Existing ceiling	Upgrade as necessary to provide two or more layers of plasterboard having staggered joints and a mass per unit area of at least 20 kg/m²

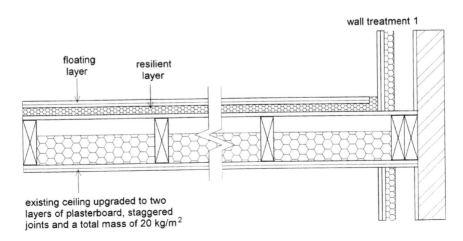

Fig. 9.3   Construction details of Floor Treatment 2 showing junction with wall treatment 1.

9.5   Stair treatment: stair covering and independent ceiling with absorbent material   AD E: 4.35–4.38

This treatment consists of adding a soft layer over the treads, which reduces impact noise at source, and constructing an independent ceiling lined with absorbent material beneath the stairs. It is the mass of the stair and independent ceiling, the isolation of the independent ceiling and the presence of the absorbent material which provides sound insulation. This is improved by constructing a cupboard enclosure beneath the stairs. It may well be necessary to consider the influence of stairs when constructing dwellings by material change of use since when stairs provide a separating function they must make the same contribution to sound insulation as separating floors.

    The specification of the stair treatment is as described in Table 9.4 and its construction form, together with associated treatment of the space beneath the stair as recommended in the AD, is shown in Fig. 9.4. Where a staircase performs a separating function, reference should be made to Building Regulation Part B – Fire Safety.

Table 9.4   Construction form of stair treatment.

	Description
Construction form	The addition of a soft layer over the existing treads and an independent ceiling lined with absorbent material beneath the stairs. A cupboard enclosure may be provided beneath the stairs
Soft covering	At least 6 mm thick, the soft covering must be securely fixed in order that it does not become a safety hazard
If there is an under stairs cupboard	The stair within the cupboard should be lined with plasterboard which has a mass per unit area of at least 10 kg/m^2. This should be mounted on battens, the space between the battens being filled with mineral wool with a density of at least 10 kg/m^3. See Fig. 9.4
	The cupboard should be built with a small heavy well fitted door and walls consisting of two layers of plasterboard, or its equivalent, each layer of which has a mass per unit area of 10 kg/m^2
If there is NOT an under stairs cupboard	In this case an independent ceiling as recommended in Floor Treatment 1 should be constructed beneath the stair

9.6   Specific junction requirements in the event of change of use AD E: 4.39–4.50

- Advice is provided in the approved document for abutments of floor treatments:
  - (a)   In the case of floating floors the resilient layer should be turned vertically upwards at room edges to ensure that there is no contact between the floating layer and the wall, and a gap of about 5 mm, filled with flexible sealant, should be left between the floating layer and the skirting.

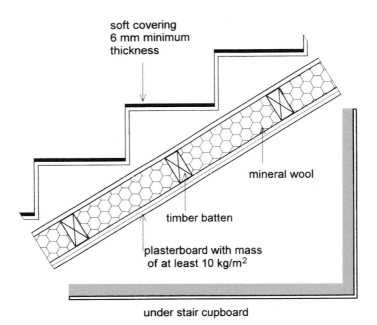

soft covering
6 mm minimum
thickness

mineral wool

timber batten

plasterboard with mass
of at least 10 kg/m^2

under stair cupboard

Fig. 9.4 Stair treatment assuming an under stair cupboard.

(b)   The junctions between new ceilings and walls should be sealed with tape or caulked with appropriate sealant, and detailed as indicated in Figs. 9.2 and 9.3.

- Junctions of separating walls and floors with other elements, e.g. external walls or other floors and walls, are a potential source of significant flanking transmission. As stated in the approved document, it may be necessary in such circumstances to seek expert advice regarding diagnosis and control of flanking transmission.

  As also stated in the approved document, if there is significant flanking transmission via adjoining walls, this can be reduced by lining all adjoining masonry walls with:

  (a)   an independent layer of plasterboard; or

  (b)   a laminate of plasterboard and mineral wool.

  The approved document suggests seeking manufacturers' advice for other dry-lining laminates and does not provide guidance as to the thickness of plasterboard which is appropriate since this will depend on the structure of the adjoining wall. Indeed, the document states that if the adjoining masonry wall has a mass per unit area of more than 375 kg/m^2, the lining may not significantly improve the insulation and so its use may not be appropriate.

- The requirements in the AD for penetrating services, i.e. piped services and ducts, passing through separating floors between flats in conversion buildings are essentially the same as for penetrating services in new buildings which are explained in section 8.1.1.

Table 9.5   Construction procedures relating to material change of use.

Element	Correct procedure	Procedures which must be avoided
Wall Treatment 1	Ensure independent panel (and frame if present) have no contact with existing wall Seal perimeter of independent panel with tape or sealant	Do not tightly compress absorbent material as this may bridge the cavity
Floor Treatments 1 and 2	Apply appropriate remedial work to the existing construction Seal perimeter of any new or independent ceiling with tape or sealant	
Floor Treatment 1		Do not create a rigid or direct connection between an independent ceiling and its floor base Do not tightly compress the absorbent material as doing so may bridge the cavity
Floor Treatment 2	Use correct density of resilient layer and ensure it can carry anticipated load Allow for movement of materials, e.g. expansion of chipboard after laying, to maintain isolation Carry resilient layer up to room edges to isolate floating layer from wall surfaces Leave a gap of approximately 5 mm between skirting and floating layer and fill with flexible sealant Lay resilient layers in sheets with joints tightly butted and taped	Do not bridge between the floating layer and the timber base or surrounding walls, e.g. with services or fixings which penetrate the resilient layer between the floating layer and timber base or walls
Junctions with floor penetrations	Seal joints between casings and ceilings with tape or sealant Leave a gap of approximately 5 mm between casing and floating layer and fill with flexible sealant.	

9.7   Construction procedures

The AD provides information regarding correct and incorrect construction procedures which will influence the sound insulation provided when undertaking material change of use. This is summarised in Table 9.5.

10 Internal Walls and Floors for New Buildings

Approved Document E, Section 5: Internal walls and floors for new buildings

10.1 General requirements   AD E: 5.1, 5.2 & 5.9–5.12

Requirement E2 of Schedule 1 to the Building Regulations 2000 (as amended) relates to protection against noise *within* dwelling places. Unlike Requirement E1 which relates to protection against noise *between* dwelling places, the performance of components within dwelling places is not assessed by pre-completion testing but simply by demonstrating that they meet the laboratory sound insulation values which are tabulated in Table 2 of Section 0: Performance of the AD, see section 5.2 of this book. This requirement relates only to airborne sound insulation and advice to occupiers wishing to improve impact sound insulation should be to provide carpets or other soft coverings to floor surfaces. The examples in the AD are for guidance, they are not exhaustive and other designs, materials or products may be used to provide the desired performance.

The following points should be considered when designing internal walls and floors:

- If a door assembly in an internal wall offers a lower level of insulation than the wall in which it is located, this will reduce the overall insulation offered. The avoidance of air paths by good perimeter sealing is an essential way of reducing sound transmission, and in addition the AD recommends the use of door sets.
- Stairs provide an important route of sound transmission between floors within dwellings, and if not enclosed the overall airborne sound insulation will be such that the potential of the associated floor will not be realised. However, as stated in the AD, the floor must still be constructed to comply with Requirement E2.
- The requirements for internal noise control should be borne in mind by designers who ought to ensure that noise sensitive rooms such as bedrooms are not constructed immediately adjacent to noise source rooms. The AD refers designers to BS 8233:1999 Sound Insulation and Noise Reduction for buildings – Code of Practice.
- The sealing of walls, floors and gaps around doors has the potential to reduce air supply and in this context the requirements of Building Regulation Part F – Ventilation and Building Regulation Part J – Combustion Appliances and fuel storage systems must be taken into account.

10.2   Internal walls   AD E: 5.4, 5.5 & 5.8

Four examples of different types of construction which should meet the requirements for internal walls are described in the AD. These are presented in the document, as follows, as wall Types A to D in a ranking order such that, as far as possible, the ones which give the best sound insulation appear first:

- Type A   Timber or metal frames with plasterboard linings on each side.
- Type B   Timber or metal frames with plasterboard linings on each side and absorbent infill material.
- Type C   Concrete block wall with plaster or plasterboard finish on both sides.
- Type D   Aircrete block wall with plaster or plasterboard finish on both sides.

10.2.1   Type A and B internal walls   AD E: 5.17–5.18

These types of walls are similar, not only in construction form but in the ways in which they insulate against airborne sound transmission. The mass of the panels and the provision of absorbent infill between them determines their resistance to airborne sound transmission which is also influenced by cavity width and material with which the frame is constructed. The construction of the examples described in the AD is as outlined in Table 10.1.

Table 10.1   Internal walls Type A and B.

Element	Wall Type A	Wall Type B
Frame	Timber or metal	Timber or metal
Linings	Linings to be provided on each side of the frame	Linings to be provided on each side of the frame
	Each lining to consist of two or more layers of plasterboard, each sheet of which has a mass per unit area of $10\,kg/m^2$ or more	Each lining to consist of a single layer of plasterboard which has a mass per unit area of $10\,kg/m^2$ or more
Distance between linings	A minimum of 75 mm if fixed to a timber frame	A minimum of 75 mm if fixed to a timber frame
	A minimum of 45 mm if fixed to a metal frame	A minimum of 45 mm if fixed to a metal frame
Absorbent layer	Not required	Unfaced wool batts or quilt with: • thickness of 25 mm or more; and • density of $10\,kg/m^3$ or more suspended in the cavity The absorbent layer may be wire reinforced
Other requirements	All joints to be well sealed	All joints to be well sealed.

10.2.2   Type C and D internal walls   AD E: 5.19–5.20

These types are similar, not only in construction form but in the ways in which they insulate against airborne sound transmission. The mass of the panels determines their resistance to airborne sound transmission. The construction of the examples described in the approved document is as outlined in Table 10.2.

Table 10.2   Internal walls Type C and D.

Element	Wall Type C	Wall Type D
Core structure	Concrete block wall	Aircrete block wall
Required mass per unit area	A mass per unit area, excluding finish, of 120 kg/m^2 or more	A mass per unit area, including finish, of <ul><li>90 kg/m^2 or more for plaster finish</li><li>75 kg/m^2 or more for plasterboard finish</li></ul>
Surface finish	Plaster or plasterboard on both sides	Plaster or plasterboard on both sides
Other requirements	All joints to be well sealed	All joints to be well sealed
Restrictions	No specific restrictions	This type of wall should:<ul><li>not be used as a load-bearing wall</li><li>not be rigidly connected to the separating floors described in the AD (see guidance relating to separating floors)</li><li>only be used with the separating walls described in the document where there is no minimum mass requirement on internal masonry walls, e.g. walls type 2.3 and 2.4 when there is no separating floor (see guidance relating to separating walls)</li></ul>

10.3   Internal floors   AD E: 5.6–5.8

Three examples of different types of construction which should meet the requirements for internal floors are described in the AD. These are presented in the document, as follows, as floor Types A to C in a ranking order such that, as far as possible, the ones which give the best sound insulation appear first:

- Type A   Concrete planks.
- Type B   Concrete beams with infilling blocks.
- Type C   Timber or metal joists with board and plasterboard surfaces.

10.3.1   Type A and B internal floors   AD E: 5.21–5.22

These types of floors are similar, not only in construction form but in the ways in which they insulate against airborne sound transmission. The mass of the planks (Type A) or beams and infilling blocks (Type B) and screed determines their resistance to airborne sound transmission. The provision of a soft covering, such as carpet, will improve impact sound insulation by reducing impact noise at its source. The construction of the examples described in the AD is as outlined in Table 10.3 and shown in Fig. 10.1.

Table 10.3   Internal floors Type A and B.

Element	Floor Type A	Floor Type B
Structure	Concrete planks	Concrete beams with infilling blocks, bonded screed and ceiling
Required mass per unit area	Mass per unit area of 180 kg/m^2 or more	Mass per unit area of concrete beams and blocks to be 220 kg/m^2 or more
Screed	The provision of a regulating screed is optional	A bonded screed is required. If a sand and cement screed is used, it should have minimum thickness of 40 mm. If a proprietary product is used, manufacturers' advice should be sought regarding its thickness
Ceiling finish	The provision of a ceiling finish is optional	Ceiling finish C or better, as defined in section 3 of the approved document, is required. See section 8.1.2 of this book for details
Other requirements	Although not stated in the approved document, floor joints should be fully grouted to ensure air tightness	

10.3.2   Type C internal floor   AD E: 5.23

Resistance to airborne sound transmission is determined by the joist and board construction, the ceiling and the absorbent material used. The provision of a soft covering, such as carpet, will improve impact sound insulation by reducing impact noise at its source. The construction form of the example described in the document is as outlined in Table 10.4 and shown in Fig. 10.1.

10.4   Internal wall and floor junctions   AD E: 5.13–5.16

Guidance on the form of junctions between separating walls and internal floors, and separating floors and internal walls is provided in sections 2 and 3 respectively of the approved document and this is considered in Chapters 7 and 8 of

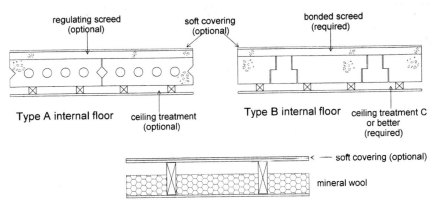

Fig. 10.1   Internal floors Type A, B and C.

Table 10.4   Internal floor Type C.

Element	Floor Type C
Structure	Timber or metal joists supporting timber or wood based boarding, a plasterboard ceiling and absorbent material between the joists
Required mass per unit area	The timber or wood based boarding should have a mass per unit area of not less than 15 kg/m^2
Ceiling finish	A ceiling consisting of a single layer of plasterboard. Mass per unit area of not less than 10 kg/m^2. Normal method of fixing
Absorbent layer	Mineral wool laid in cavity between joists. Thickness not less than 100 mm and density not less than 10 kg/m^3
Other requirements	See BRE BR 262 *Thermal Insulation: Avoiding risks*, section 2.4 [17] regarding heat emission from electrical cables which may be covered by absorbent material

this book. When separating elements are present, the junctions between them and internal walls and floors should always be constructed in accordance with this guidance.

Whenever internal walls or floors are constructed, care should be taken to ensure that there are no air paths, i.e. direct or indirect routes of air passage, between rooms by filling all gaps around the wall or floor.

11 Rooms for Residential Purposes

Approved Document E, Section 6: Rooms for residential purposes

11.1 General requirements AD E: 6.1–6.3 & 6.16–6.18

The expression 'rooms for residential purposes', which is defined in Regulation 2, is used to refer to rooms such as those in hotels, hostels, boarding houses, halls of residence and residential homes but not those in, for example, hospitals.

Section 6 of the Approved Document E gives examples of walls and floors which it states should achieve the performance standards laid out in section 0: Performance – Table 1b of the AD and Chapter 5 of this book although, as is stated in the document, the actual performance of a wall or floor will depend on the quality of its construction. The information in this section of the AD is provided only for guidance. It is not exhaustive and the details in no way override the requirement to undergo pre-completion testing. Designers are recommended also to seek advice from other sources such as manufacturers regarding alternative designs, materials or products.

It is not only sound insulation but room layout and building services which determine internal noise levels and these factors may be particularly significant in the confined spaces of 'rooms for residential purposes'. The requirements for internal noise control should be borne in mind by designers who ought to endeavour to ensure that noise sensitive rooms such as bedrooms are not constructed immediately adjacent to noise source rooms. The AD refers designers to BS 8233:1999 *Sound insulation and noise reduction for buildings – code of practice* and the BRE/ CIRIA Report: *Sound control for homes* [12].

11.2 New buildings containing rooms for residential purposes

11.2.1 Separating walls AD E: 6.4

Examples of alternative types of separating walls suitable for new buildings are presented in section 2 of the AD and described in Chapter 7 of this book. It is stated in the AD that where buildings contain rooms for residential purposes, of the examples given, the most appropriate choices are:

- Wall Type 1 (solid masonry), where the in situ plastered finishes associated with each of the variants 1.1, 1.2 and 1.3 may be substituted for by a single sheet of plasterboard on each face provided that the sheets each have a mass per unit area of $10\,kg/m^2$ or more

- Wall Type 3 (masonry between independent panels) but only Types 3.1 and 3.2 which have solid masonry cores.

The AD states that wall Type 2 (cavity masonry) and wall Type 4 (framed wall with absorbent material) may be used but particular care must be taken to maintain isolation between the leaves and specialist advice may be required.

11.2.2   Separating floors   AD E: 6.8

Examples of alternative types of separating floors suitable for new buildings are presented in section 3 of the approved document and described in Chapter 8 of this book. It is stated in the AD that where the buildings contain rooms for residential purposes, of the examples given, the most appropriate choice is one of the following subgroups of floor Type 1 (concrete base with soft covering):

- Floor Type 1.1C   Solid concrete slab, cast in-situ with or without permanent shuttering. Soft floor covering and ceiling treatment C
- Floor Type 1.2B   Hollow or solid concrete planks. Soft floor covering and ceiling treatment B.

The AD states that floor Type 2 (concrete base with ceiling and floating floor) and floor Type 3 (timber frame base with ceiling and platform floor) may be used but their floating floors and ceilings must not be continuous across the walls which separate rooms for residential purposes. Designers are advised that this type of construction may require specialist advice.

11.3   Corridor walls and doors   AD E: 6.5–6.7

The AD states that the walls between rooms for residential purposes and corridors should be built to the specification of the separating walls, and their junction details, recommended for use between rooms for residential purposes in Approved Document E section 6; see also this chapter. However, a weak point in the sound insulation provided from corridor noise is that transmitted through doors. For this reason, the document states that doors to corridors should have good sealing around their perimeters (including, if practical, their thresholds), and a mass per unit area of at least $25\,kg/m^2$, or alternatively use a doorset with a weighted sound reduction index of at least 29 dB. The term sound reduction index is described in Chapter 3 and the relevant measurement standards are listed in section 6.9.3.3.

Particular attention should be paid to potentially noisy places such as bars, which ideally should be separated from the rest of the building by a lobby, two doors in series (i.e. such that each door is passed through in turn) or a high performance doorset. If this provision cannot be made to the noisy room, it should be applied to rooms for residential purposes in its vicinity.

The requirements of Building Regulation Part B – Fire safety, and Building Regulation Part M – Access and facilities for disabled people, must be taken into account when selecting doors for buildings of this type.

11.4 Material change of use – rooms for residential purposes AD E: 6.9–6.10

Section 6 of the AD considers forms of treatment to walls, floors, stairs and their associated junctions which it may be appropriate to apply when converting existing buildings by material change of use to provide rooms for residential purposes. The point is made in the AD that existing components may already satisfy the requirements of Table 1b of section 0: Performance (see Chapter 5 of this book) and it is suggested that this would be so if the construction of a component, its junction detailing and flanking construction were sufficiently similar to those of one of the example walls or floors described above for 'rooms for residential purposes' created in new buildings. It is suggested in the AD that, with respect to requirements for mass, for example, an existing component may be sufficiently similar to comply with the required performance requirements if its mass per unit area is within 15% of that of an equivalent component recommended for the construction of new buildings. This may be the case, but due to uncertainty about the workmanship and density of material used throughout the component, additional treatment may still be advisable.

The construction of an existing component may be such that it is not possible to demonstrate whether or not it will comply with the requirements of Table 1b of section 0: Performance. In this case, use may be made of the floor, wall and stair treatments recommended in section 4 of the AD for houses and flats resulting from change of use, see Chapter 9 of this book. The design of separating components and their associated junctions to provide adequate sound insulation in such circumstances may be of a complex nature and developers are advised in the approved document that specialist advice may be required when undertaking this type of work.

11.5 Junctions   AD E: 6.11–6.15

As with other forms of dwelling place, in order for separating elements to fulfil their full potential it is essential that flanking transmission is restricted by careful junction design and selection of flanking elements. The AD states that in the case of new buildings, the guidance relating to junction design and flanking transmission provided in sections 2 and 3 of the document for separating walls and floors respectively of houses and flats should be adhered to when building 'rooms for residential purposes'. This guidance is described in Chapters 7 and 8 of this book.

Similarly, the guidance relating to junction and flanking details for houses and flats formed by material change of use in section 4 of the AD should be applied when creating 'rooms for residential purposes' by material change of use. This guidance is explained in Chapter 9 of this book.

A relaxation to the guidance for houses and flats is that where a solid masonry Type 1 separating wall is used, the wall need not be continuous through a ceiling void or roof space to the underside of a structural floor or roof, subject to the following conditions:

- There is a ceiling consisting of two or more layers of plasterboard with a total mass per unit area of not less than 20 kg/m^2
- There is a layer of mineral wool in the roof void which is not less than 200 mm thick and has a density of not less than 10 kg/m^3; and
- The ceiling is not perforated.

Also, the structure of the building should be such that neither the ceiling joists nor plasterboard sheets are continuous across the wall separating 'rooms for residential purposes' with sealed joints as shown in Fig. 11.1. As stated in the approved document, this construction detail may only be used if Building Regulation Part B – Fire Safety, and Building Regulation Part L – Conservation of Fuel and Power, are satisfied.

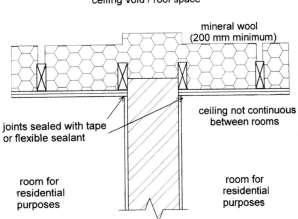

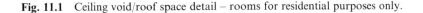

Fig. 11.1 Ceiling void/roof space detail – rooms for residential purposes only.

12 Reverberation in Common Parts of Buildings

Approved Document E, Section 7: Reverberation in the common internal parts of buildings containing flats or rooms for residential purposes

12.1 General requirements AD E: 7.1–7.9

The common parts of buildings containing flats and/or 'rooms for residential purposes' tend to be constructed with hard durable surface finishes, which are easily maintained. Unfortunately, such surfaces lack the soft open texture which efficiently absorbs sound and so the level of reflected, or reverberated, sound tends to be high in such places. Requirement E3 of Part E of the Building Regulations 2000 (as amended) states that the design and construction of the common parts of such buildings shall be such as to prevent more reverberation around them than is reasonable.

Fortunately, it is relatively easy to increase sound absorption and hence reduce reverberant noise levels by surface treatment. Procedures for determining the amount of additional absorption which is required in corridors, hallways, stairwells and entrance halls providing access to flats and 'rooms for residential purposes' places are described in section 7 of Approved Document E. The document differentiates between different types of common space as shown in Table 12.1.

Table 12.1 Limiting dimensions of common areas.

Type of space	Ratio of longest to shortest floor dimensions
corridor, hallway	greater than three
entrance hall	three or less
stairwell	not defined in terms of ratio of dimensions

The types of space referred to in Table 12.1 are frequently interconnected and in such circumstance the guidance for each space should be followed individually.

The AD describes two methods, A and B, for calculating how much sound absorption is required and the procedure for the use of each method is explained in sections 12.3 and 12.4 with an example of their application in section 12.5. However, whichever procedure is adopted, it is essential that any material chosen to line the internal surfaces complies with the requirements of Building Regulation Part B – Fire Safety and it is recommended in the AD that if designers require guidance

148

regarding the provision of additional sound absorbent material, this should be sought at an early stage of the design process.

12.2   Sound absorption properties of materials

When sound energy strikes a surface, some passes through it, some is absorbed by it and some is reflected. The principles are discussed in Chapter 2, where the definition of 'the sound absorption coefficient', α, is given.

If no sound is reflected $\alpha = 1$ and if none is absorbed or transmitted $\alpha = 0$. In theory, all building materials have absorption coefficients between zero and one, values close to zero representing reflective surfaces with low absorption power and vice versa. Sound absorption is highly dependent on frequency. Some of the types of material used to line rooms have good absorption power in the mid to high frequencies of the audible spectrum whereas for others it is highest at the lower end of the spectrum. For this reason it is usual to quote one-third octave or one octave band values of absorption coefficients measured across the audible spectrum. Octave-band values of absorption coefficient for a range of materials are shown in Table 12.4 below.

12.3   Calculation of additional absorption – Method A
AD E: 7.10–7.12

This method makes use of a procedure for classifying sound absorbers which is described in EN ISO 11654:1997 . To apply this procedure, absorption coefficients are measured at one-third octave bands and the arithmetic mean of the three one-third octave bands within each octave band is taken to obtain the 'practical sound absorption coefficient', α_{pi}, a one octave band average value where i is the ith octave band. The one octave band values are then compared with a reference curve which is 'shifted' in increments of 0.05dB until the sum of the adverse deviations does not exceed 0.01dB. The value of the reference curve at 500Hz, α_w, is then taken as a single figure weighted average absorption coefficient of the material in question. The material is then given a classification letter according to its value of α_w as shown in Table 12.2 which is an extract from EN ISO 11654:1997.

Method A involves covering a prescribed area with an absorbent material of a defined classification as follows.

12.3.1   Entrance halls, corridors, hallways

Cover an area, at least as great as the floor area, with an absorber which is Class C or better. The document states that it will normally be expedient to cover the ceiling with the additional absorptive material.

Table 12.2 Sound absorption classes defined in EN ISO 11654:1997.

Sound absorption class	α_w
A	0.90; 0.95; 1.00
B	0.80; 0.85
C	0.60; 0.65; 0.70; 0.75
D	0.30; 0.35; 0.40; 0.45; 0.50; 0.55
E	0.25; 0.20; 0.15
Not classified	0.10; 0.05; 0.00

12.3.2   Stairwells and stair enclosures

Calculate the combined area of:

- stair treads
- upper surface of intermediate landings
- upper surface of landings (excluding ground floor); and
- ceiling area on the top floor

and cover an area which is equal to or larger than the calculated area with a Class D absorber, or cover an area which is at least half of the calculated area with an absorber of Class C or better. In either case the sound absorbing material should be distributed equally between floor levels. It is stated in the approved document that it will normally be appropriate to cover the:

- underside of intermediate landings
- underside of the other landings; and
- ceiling area on the top floor.

The materials used to increase sound absorption tend to be light and porous, lacking the robustness which is required to withstand day to day wear and tear and vandalism in common areas where they may come into contact with human beings. This why it is recommended in the AD that they are used to cover ceilings and the undersides of landings despite the fact that they would be equally effective on other surfaces. The AD states that the use of proprietary acoustic ceilings is appropriate. These have the benefits of being available with a variety of surface finishes and are lightweight and very efficient absorbers of sound.

12.4   Calculation of additional absorption – Method B
AD E: 7.13–7.21

This method takes into account the actual absorption power of the surfaces of the enclosure prior to the provision of additional absorbent material and allows the amount of additional material which is required to be calculated. It is stated in the approved document that it is only intended for corridors, hallways and entrance halls (not stairwells) but that it offers a more flexible approach which may require less additional absorption than Method A. It is also a more efficient approach in that the additional absorption can be directed at the frequencies at which it is most needed.

For a particular surface of area S m^2 and absorption coefficient α, its absorption area A is defined as the product of S and α and it has units of m^2. This may be considered as the hypothetical area of a material which is a perfect absorber, i.e. one with $\alpha = 1$, which would provide the same sound absorption as the actual material with area S.

If an enclosure has n internal surfaces, its total absorption area, A_T is defined as the sum of the absorption areas of its individual components, such that:

$$A_T = \alpha_1 \, S_1 + \alpha_2 \, S_2 + \alpha_3 \, S_3 + \ldots\ldots + \alpha_n \, S_n$$

It follows that A_T may be considered as the hypothetical area of a material which is a perfect absorber which would provide the same total sound absorption as the sum of the actual materials which are present.

Requirement E3 will be satisfied when the total absorption area stipulated in Table 12.3 for different types of common area has been provided at each octave band from 250Hz to 4000Hz.

Table 12.3 Total absorption areas for common areas using Method B.

Type of common area	Minimum total absorption area required	Location of additional absorptive material
Entrance halls	0.20 m^2 per cubic metre of volume	To be distributed over the available surfaces
Corridors, hallways	0.25 m^2 per cubic metre of volume	To be distributed over one or more of the surfaces

Absorption coefficient data which should be accurate to two decimal places may be obtained from a variety of sources. For generic materials, data is provided in Table 7.1 of the AD which is reproduced as Table 12.4. It is stated in the AD that this may be supplemented by other published data. Whilst published data will be necessary and appropriate for generic surface materials not provided specifically to increase the absorption area of the enclosure, materials which are selected for their absorption properties will usually be trade products and their absorption coefficients should be available from their manufacturers or suppliers. These should be established by measuring one-third octave band values of absorption coefficients in accordance with BS EN 20354:1993 and then converting them to one octave band values of 'practical sound absorption coefficient', α_{pi}, using the procedure described in EN ISO 11654:1997, see section 12.3.

Table 12.4   Copy of Table 7.1 from Approved Document E providing absorption coefficients for commonly used materials.

Material	Sound absorption coefficient, in octave frequency bands (Hz)				
	250	500	1000	2000	4000
Fair-faced concrete or plastered masonry	0.01	0.01	0.02	0.02	0.03
Fair-faced brick	0.02	0.03	0.04	0.05	0.07
Painted concrete block	0.05	0.06	0.07	0.09	0.08
Windows, glass facade	0.08	0.05	0.04	0.03	0.02
Doors (timber)	0.10	0.08	0.08	0.08	0.08
Glazed tile/marble	0.01	0.01	0.01	0.02	0.02
Hard floor coverings (e.g. lino, parquet) on concrete floor	0.03	0.04	0.05	0.05	0.06
Soft floor coverings (e.g. carpet) on concrete floor	0.03	0.06	0.15	0.30	0.40
Suspended plaster or plasterboard ceiling (with large airspace behind)	0.15	0.10	0.05	0.05	0.05

When calculating the area and absorption coefficient of the material necessary to satisfy the requirements of Table 12.3, calculation steps should be rounded to two decimal places.

12.5   Example of absorption calculation

There follows an example of the application of Methods A and B to calculate the extra absorption required for an entrance hall, see Fig. 12.1.

12.5.1   Application of Method A

This requires covering an area equal or greater than the floor area with a Class C, or better, absorber. An appropriate solution would therefore be to cover $4.0 \times 5.0 = 20 \, m^2$, e.g. the ceiling with an appropriate material, probably acoustic ceiling tiles, of Class C or better.

12.5.2   Application of Method B

Assuming that the designer has decided to cover the entire ceiling with absorptive material, the necessary calculations are as described below.

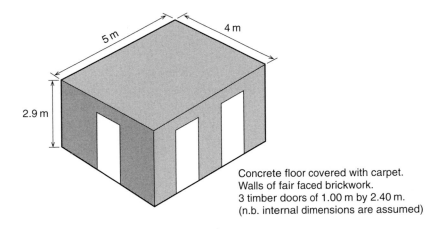

Concrete floor covered with carpet.
Walls of fair faced brickwork.
3 timber doors of 1.00 m by 2.40 m.
(n.b. internal dimensions are assumed)

Fig. 12.1   Entrance hall relating to example question.

Stage 1

Calculate area of each internal surface and establish the absorption coefficient from Table 12.4 for each surface which is not being selected for its sound absorbing properties.

Surface	area (m²)	Absorption coefficient for each octave band				
		250 (Hz)	500 (Hz)	1000 (Hz)	2000 (Hz)	4000 (Hz)
Floor (wood, carpet covered)	20	0.03	0.06	0.15	0.30	0.40
Doors (timber)	7.2	0.10	0.08	0.08	0.08	0.08
Walls (exc. doors) (fair-faced brick)	45	0.02	0.03	0.04	0.05	0.07
Ceiling (yet to be determined)	20					

Note that the carpeted floor provides good absorption at the higher frequency octave bands. This reduces the absorption required at these frequencies from the ceiling covering.

Stage 2

Calculate the absorption area of floor, doors and walls (the product of area and absorption coefficient) and then establish the overall absorption area at each octave band of these surfaces by summing:

Surface	Absorption area (m²)				
	250 (Hz)	500 (Hz)	1000 (Hz)	2000 (Hz)	4000 (Hz)
Floor	0.60	1.20	3.00	6.00	8.00
Doors	0.72	0.58	0.58	0.58	0.58
Walls (exc. doors)	0.90	1.35	1.80	2.25	3.15
Sum of absorption areas	2.22	3.13	5.38	8.83	11.73

Stage 3
Calculate the total absorption area, A_T, required at each octave band, i.e. $0.2 \times$ volume $= 0.2 \times 5 \times 4 \times 2.9 = 11.6 \text{ m}^2$

Stage 4
Calculate the additional absorption area required from ceiling by subtracting the values found at stage 2 from that found at stage 3:

	Absorption area (m²)				
	250 (Hz)	500 (Hz)	1000 (Hz)	2000 (Hz)	4000 (Hz)
Additional area required from ceiling	9.38	8.47	6.22	2.77	−0.13

The negative value at 4000Hz indicates that sufficient absorption is provided at this octave band by the floor, doors and walls.

Stage 5
Calculate the absorption coefficient required for the ceiling by dividing the required absorption area by the area of the ceiling:

	Absorption area (m²)				
	250 (Hz)	500 (Hz)	1000 (Hz)	2000 (Hz)	4000 (Hz)
Required absorption coefficient of ceiling material	0.47	0.42	0.31	0.14	any value

   Select a patent ceiling product which has absorption coefficients greater than those in the above table.

12.6   Report on compliance   AD E: 7.22

Demonstration of compliance with Requirement E3 may take the form of an annotated drawing or a report. The information which should be included, quoted directly from the approved document, is as follows:

(1)   A description of the enclosed space (entrance hall, corridor, stairwell etc.)
(2)   The approach used to satisfy Requirement E3, Mcthod A or B:
  •   With Method A, state the absorber class and the area to be covered.
  •   With Method B, state the total absorption area of additional absorptive material used to satisfy the Requirement.
(3)   Plans indicating the assignment of the absorptive material in the enclosed space.

13 Robust Standard Details

An alternative to pre-completion testing

13.1 Introduction

Requirement E1 of Schedule 1 to the Building Regulations 2000 stipulates that a predefined level of acoustic performance must be provided between adjoining dwelling places and, via Regulations 12A and 20A, that this should be demonstrated by way of a programme of pre-completion testing, as described in Chapter 6. This procedure ensures that buildings are constructed not only to a high notional standard of sound insulation but also that compliance with the standard will be verified by routine testing of a sample of completed buildings. The pre-completion requirement is particularly attractive to potential purchasers, because it provides positive evidence that the property meets the legal standard for sound insulation. However the scheme is less appealing to the house building industry for a number of reasons:

- The pre-completion programme introduces an additional stage into the construction process
- There are extra costs and possible delays in completion and hand-over
- A failed test can have serious consequences in terms of remedial work
- Remedial work and its associated costs and delays may extend to other properties in addition to the one that failed.

Clearly, the consequences of all these disadvantages may be quite significant for the builder, and it can also be argued that some or even all of the extra costs will eventually be borne by the purchaser. In response to this, the House Builders Federation have proposed an alternative solution which is designed to avoid the potential disadvantages of routine pre-completion testing.

13.2 Robust standard details

The principal reason for introducing pre-completion testing was that for various reasons, such as design detailing and construction site practices, components have not in the past been achieving their full performance potential. It follows that if there could be a sufficiently high level of confidence that the performance levels specified in the Building Regulations were being achieved in practice, such that the pass rate in pre-completion testing approached 100%, the need for the pre-completion test programme would be reduced.

With this in mind, the House Builders Federation, which represents the interests of a large number of the nation's house builders, have proposed the introduction of a series of robust standard details (RSDs) for use as separating walls and separating floors which would not require routine pre-completion testing. The justification which is proposed for not applying the pre-completion tests is that the RSDs would be designed such that the level of insulation provided would be consistently greater than the standard required to satisfy requirement E1.

To guarantee that an installed separating wall or floor will provide a performance, in terms of sound insulation, which is greater than the prescribed values of $D_{nT,w} + C_{tr}$ in Approved Document E, is difficult since there are inherent variations in the composition of the materials forming the components as well as quality of workmanship. Due allowance has had to be made for variables such as these in selecting target values for the sound insulation offered by an RSD.

It follows that in order to be assured that virtually all separating components comply with the legislative requirement, the relevant average, or design, performance value will have to be considerably better than the performance criteria prescribed in the approved document. Thus, in those properties where, due to the inherent and inevitable variability of building work, the performance falls below the design value, the sound insulation will still be sufficient to comply with the minimum standard set in the approved document. The proponents of this approach consider it to have potential benefits to both the client (building occupier) and the builder. For the building occupier, the expected benefit is that in the majority of dwellings the standard of sound insulation will be significantly higher than that prescribed by the approved document. For the builder, the cost of constructing to a higher standard of insulation may well be higher than working to approved document guidance but in return the requirement for pre-completion testing, and its attendant uncertainty, will be avoided, resulting in a possible cost saving. A further advantage of using RSDs may accrue to developers who will be able to claim that the properties which they are marketing offer a notional level of sound insulation between separating elements which is, on average, greater than that required under current Building Regulations.

13.3   The House Builders Federation robust standard details project

Shortly before the publication of Approved Document E: 2003, a project was initiated by the House Builders Federation targeted at providing a series of RSDs for new houses and apartments as an alternative to pre-completion testing. The impetus for this work was a ministerial statement to the effect that the Building Regulations Advisory Committee will give consideration as to whether or not the RSDs produced by the House Builders Federation achieve their objective.

The project was undertaken between July 2002 and July 2003 under the management of The Building Performance Centre at Napier University. Its formal objectives, as itemised on the House Builders Federation website, were to:

- Provide consistent levels of performance that exceed the performance standards given in Approved Document E
- Be acceptable to the Building Regulations Advisory Committee, the Office of the Deputy Prime Minister, House Builders Federation members and industry
- Include a safety margin to allow for normal workmanship variance
- Reduce the effect of design/material weaknesses that influence the results
- Include a broad range of industry products and practices
- Provide sufficient clarity and instruction for correct implementation on site
- Be compatible with other regulations. (e.g. Part L and Health and Safety).

Compliance with the first, third and fourth of the above items, which are inter-related, is intended to ensure that the required legal standards are met and in this context the provision of adequate instructions for site implementation is imperative and correctly stipulated as an objective. The inclusion of a broad range of products and practices within the available RSDs is important in order that an adequate choice between alternatives is available to both house builders and house buyers. Due to the complexity and interrelationships between Building Regulations and also between them and other regulations, it is essential that issues relating to compat-ibility with other regulations, referred to in the last item listed above, are fully addressed when drafting requirements relating to particular facets of construction. Clearly, the proposed details must be acceptable to all the parties referred to under the second item above and if that is achieved then the House Builders Federation proposal should produce successful solutions.

The project was structured by setting up working groups relating to the following aspects of construction:

- masonry and concrete
- timber
- steel

together with a group to deal with issues relating to regulations and an overseeing steering committee. The working groups, which included representatives from construction interests ranging from house builders to acoustics experts, have examined proposed (or candidate) RSDs and a comprehensive testing programme has been undertaken by a series of measurement contractors in order that con-structions worthy of RSD status may be selected. The result of this work was pre-sented to the Building Regulations Advisory Committee during the summer of 2003, and a consultation document, 'Amendment of the Building Regulations to allow Robust Standard Details to be used as an alternative to pre-completion testing' was issued in August 2003.

In this scheme, it is proposed that to qualify for the status of robust standard detail, that detail must have been constructed and tested at least 30 times, and must have exceeded the performance targets defined in AD E by a specified margin. For the airborne sound insulation of separating floors, the results for $D_{nT,w} + C_{tr}$ must exceed the approved document minimum of 45dB as follows:

- The mean of all 30 tests must exceed the AD E value by at least 5dB, and
- The lowest result within the 30 tests must exceed the AD E value by at least 2dB.

For the impact sound insulation of separating floors, $L'_{nT,w}$ must be below the approved document maximum of 62dB as follows:

- The mean of all 30 tests must be below the AD E value by at least 5dB, and
- The highest result within the 30 tests must be below the AD E value by at least 2dB.

In operation, the proposed RSD scheme would have the following components:

- A management board and secretariat
- A complaints procedure
- RSD specification sheets and checklists
- A method for monitoring the performance of previously approved RSDs.

The monitoring procedure has not been defined. One possibility is that RSDs should be subjected to the same pre-completion testing procedures as other constructions but with a lower rate of testing.

Before the RSD scheme can be adopted, it is likely that some amendments to the Building Regulations 2000 will be necessary. In particular, Regulations 20A and 12A may have to be altered. However, even if it is adopted, the RSD scheme will not be a replacement for pre-completion testing (as presently defined), but will sit alongside it as an alternative.

14 School Acoustics

Requirement E4 states that the design and construction of all rooms and other places in school buildings shall be such that their acoustic conditions and insulation against disturbance by noise are appropriate to their intended use. It is stated in Approved Document E that the normal way of satisfying these conditions will be to meet the requirements given in section 1 of Building Bulletin 93 *The acoustic design of schools* [2] with respect to sound insulation, reverberation time and internal ambient noise levels. Section 1 is only part of BB93, the complete document containing specifications, information and guidance on all aspects of the acoustic design of schools. The main sections are:

(1) Specification of acoustic performance
(2) Noise control
(3) Sound insulation
(4) The design of rooms for speech
(5) The design of rooms for music
(6) Hearing requirements
(7) Case studies.

The performance standards set out in section 1 are more extensive than those in Approved Document E, and cover the following:

- Indoor ambient noise levels in unoccupied spaces
- Airborne sound insulation between spaces
- Airborne sound insulation between circulation spaces and other spaces used by pupils
- Impact sound insulation of floors
- Reverberation in teaching and study spaces
- Sound absorption in corridors, entrance halls and stairwells
- Speech intelligibility in open-plan spaces.

Section 1 also covers demonstration of compliance to both the building control body and the client.

In practice, those concerned with school design and construction will find it necessary to consult the whole of BB93, especially as there is some cross-referencing between its various sections.

Appendix
The calculation of mass
Approved Document E, Annex A. Method for calculating mass per unit area

The mass per unit area, expressed in kilograms per square metre may be obtained from manufacturers' data. Alternatively, it can be calculated for walls and floors by the following methods as described in Annex A of Approved Document E.

A.1 Walls AD E: A2–A4

In the case of a brick or block wall, mortar beds and perp ends may constitute a considerable proportion of the total volume and, since the density of mortar may be significantly different to that of the blocks, account must be taken of this when calculating the mass per unit area. The procedure used is to define the 'co-ordinating area' as shown in Fig. A.1, and to calculate the mass per unit area, M_A from:

$$M_A = \frac{\text{mass of co-ordinating area}}{\text{co-ordinating area}} = \frac{M_B + \rho_m(T\,d\,(L + H - d) + V)}{LH}\,\text{kg/m}^2$$

where, in the above equation:

M_B = brick or block mass (kg)
ρ_m = density of mortar (kg/m^3)
T = brick or block thickness (m) (excluding any finish)
d = mortar joint thickness (m)
L = co-ordinating length (m)
H = co-ordinating height (m)
V = volume of any mortar filled void, e.g. a frog.

In using the above equation the following points should be noted.

- For application of the above equation, the brick or block unit must penetrate the full thickness (excluding finishes) of the wall.
- Since density is a function of moisture content, the mass of brick/block and density of mortar should be at taken at the appropriate moisture content. The AD makes reference to Table 3.2 of CIBSE Guide A (1999) for this information.

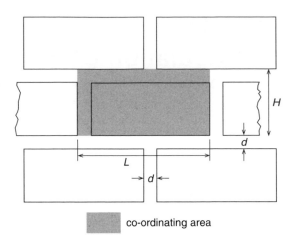

Fig. A.1   Co-ordinating area used for calculation of mass per unit area.

- The above formula calculates the mass per unit area of a single leaf without surface finish. If a finish, e.g. plaster, is to be applied and/or if it is a cavity wall with another leaf, the mass per unit area of each component must be added together to obtain the total mass per unit area of the wall.
- Manufacturers' data should be used to obtain the mass per unit area of surface finishes.

Example calculation

Assume that a concrete block has the following dimensions:

thickness   =   0.120 m
length   =   0.440 m
height   =   0.290 m
mass of block =   14 kg

and it is to be used on edge with mortar which has a density of 1800 kg/m^3 and a thickness of 0.010 m (10 mm).

In this case: MB =   14 kg
ρ_m =   1800 kg/m^3
T =   0.120 m
d =   0.010 m
L =   0.450 m
H =   0.300 m
V =   0 (it is assumed that there is no frog)

and substituting into the above equation gives a mass of 115.5 kg/m^2.

The AD makes reference to 'simplified equations'. These, however, are simply representations of the equation shown above with substitution for all variables for the particular case with the exception of the mass of the block. In this format the above equation would be written as:

$$\text{Mass per unit area} = 7.4\, M_B + 11.8\ \text{kg/m}^2$$

Various values of M_B may now be substituted into the equation to establish the mass per unit area.

A.2   Floors   AD E: A5

The mass per unit area of a flat solid concrete slab floor, M_F, may be obtained by multiplying its density, ρ_c, by its thickness, T:

$$M_F = \rho_c\, T$$

In the case of a conventional beam and block floor, as shown in Fig. A.2, the mass per unit area, M_F, can be obtained from:

$$M_F = \frac{(M_{beam,1m} + M_{block,1m})}{L_B}$$

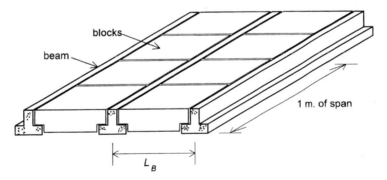

Fig. A.2   Beam and block floor.

where, in the above equation:

$M_{beam,1m}$	=	the mass of 1 m of the length of the beam
$M_{block,1m}$	=	the mass of the blocks spanning between two consecutive beams for 1 m of the length of the beam
L_B	=	the dimension defined in Fig. A.2.

No other examples are given in the AD. Designers are advised in the AD to seek advice from manufacturers on the mass per unit area of other floor types.

British Standards

BS 1243:1978
   Metal ties for cavity wall construction.

BS 5628-3:2001
   Code of practice for use of masonry – Part 3 – Materials and components, design and workmanship.

BS 5821-3:1984 ISO 717/3:1982
   Methods for rating the sound insulation in buildings and of building elements – Part 3: Method for rating the airborne sound insulation of façade elements and façades.
   [ISO title: Acoustics – Rating of sound insulation in buildings and of building elements – Part 3: Airborne sound insulation of façade elements and façades.]

BS 8233:1999
   Sound insulation and noise reduction for buildings – Code of practice.

BS EN ISO 140-1:1995 (BS 2750-3:1995)
   Acoustics – Measurement of sound insulation in buildings and of building elements – Part 1: Requirements of laboratory test facilities with suppressed flanking transmission.

BS EN ISO 140-2:1995 (BS 2750-3:1995)
   Acoustics – Measurement of sound insulation in buildings and of building elements – Part 2: Determination, verification and application of precision data.

BS EN ISO 140-3:1995 (BS 2750-3:1995)
   Acoustics – Measurement of sound insulation in buildings and of building elements – Part 3: Laboratory measurement of airborne sound insulation of building elements.